2nd International Conference on Design, Mechanical and Material Engineering (D2ME 2017)

MATEC Web of Conferences Volume 136 (2017)

Melbourne, Australia
14 - 16 September 2017

Editors:

Raj Das
Alan Kin Tak Lau

ISBN: 978-1-5108-5207-5

Printed from e-media with permission by:

Curran Associates, Inc.
57 Morehouse Lane
Red Hook, NY 12571

Some format issues inherent in the e-media version may also appear in this print version.

Printed by Curran Associates, Inc. (2017)

For additional information, please contact EDP Sciences – Web of Conferences at the address below.

EDP Sciences – Web of Conferences
17, Avenue du Hoggar
Parc d'Activité de Courtabœuf
BP 112
F-91944 Les Ulis Cedex A
France

Phone: +33 (0) 1 69 18 75 75
Fax: +33 (0) 1 69 28 84 91

contact-edps@webofconferences.org

Additional copies of this publication are available from:

Curran Associates, Inc.
57 Morehouse Lane
Red Hook, NY 12571 USA
Phone: 845-758-0400
Fax: 845-758-2633
Email: curran@proceedings.com
Web: www.proceedings.com

TABLE OF CONTENTS

PREFACE

2017 2nd International Conference on Design, Mechanical and Material Engineering (D2ME 2017) were held in Swinburne University of Technology, Melbourne, Australia, from September 14-16, 2017. The proceeding is a compilation of the accepted papers and represent an interesting outcome of the conference.

D2ME 2017 is co-organized by Swinburne University of Technology and Hong Kong Society of Mechanical Engineers, supported by Southwest Jiaotong University. The conferences provide a useful platform both for showing the latest research directions and for exchange of research results and ideas in broad Design, Mechanical and Material Engineering area.

We would like to acknowledge all of those who supported D2ME 2017. Each individual and institutional help were very important for the success of this conference. Especially we would like to thank the organizing committee for their valuable advices in the organization and helpful peer review of the papers.

We hope that D2ME 2017 will be a forum for fruitful discussions that will shed light on new ideas and foster collaborative researches. We are sure that the proceedings will serve as an important research source of references and the knowledge, which will lead to not only scientific and engineering progress but also new products and processes.

Dr Raj Das
RMIT University, Australia
Prof. Alan Kin-Tak Lau
Swinburne University of Technology, Australia

D2ME 2017 COMMITTEE

Advisory Co-Chairs

Prof. Guy Littlefair, Deakin University, Australia

General Chair

Dr. Raj Das, RMIT University, Australia

Conference Co-Chairs

Prof. Zhan Chen, AUT University, Auckland, New Zealand
Prof. LAU Kin Tak Alan, Swinburne University of Technology, Australia
Prof. Qinghua Qin, The Australian National University, Australia

Program Committee Chair

Prof. Gong Hao, National University of Singapore, Singapore

International Technical Committee

Prof. Wong Shaw Voon, Universiti Putra Malaysia, Malaysia
Prof. Li Chang, The University of Sydney, Australia
Prof. Alan C. Lin, National Taiwan University of Science and Technology, Taiwan
Dr. Abhishek Kundu, Cardiff University, UK
Dr. Khalid Arif, Massey University, New Zealand
Dr. Pierre Kerfriden, Cardiff University, UK
Dr. Chi Kit Au, The University of Waikato, New Zealand
Dr. Kheng-Lim Goh, Newcastle University, UK
Dr. Hao Wang, Northestern University, China
Dr. Prodip Das, Newcastle University, UK
Dr. Yat Wong, Swinburne University of Technology, Australia
Dr. Jie Zhang, University of Bristol, U
Dr. Zhangming Wu, Cardiff University, UK
Dr. Alokesh Pramanik , Curtin university, Australia

Dr. Pooya Sareh, Imperial College London,UK
Dr. Ching L. Ko, University of Oklahoma, USA
Dr. Mohsen Motahari-Nezhad, Zabol University, Zabol, Iran
Dr. Yang Xi, Researcher at Diodes Inc., United States
Dr. GUELAILIA Ahmed, Algerian Space Agency, Algeria.

Statement of Peer review

In submitting conference proceedings to MATEC Web of Conferences, I certify to the Publisher that I adhere to the **Policy on Publishing Integrity** of the journal in order to safeguard good scientific practice in publishing.

1. All articles have been subjected to peer review administered by the proceedings editors.
2. Reviews have been conducted by expert referees, who have been requested to provide unbiased and constructive comments aimed, whenever possible, at improving the work.
3. Proceedings editors have taken all reasonable steps to ensure the quality of the materials they publish and their decision to accept or reject a paper for publication has been based only on the merits of the work and the relevance to the journal.

Title, date and place of the conference

2017 2nd International Conference on Design, Mechanical and Material Engineering (D2ME 2017)

September 14-16, 2017 Swinburne University of Technology, Australia

Proceedings editor(s): Raj Das Alan Kin Tak LAU

Date and editor's signature 20 October 2017

Analysis of Material Removal and Surface Characteristics in Machining Multi Walled Carbon Nanotubes Filled Alumina Composites by WEDM Process

Meinam Annebushan Singh[1], Deba Kumar Sarma[2a], Hriday Mani Kalita[3] and Comingstarful Marthong[4]

[1]*NIT Meghalaya, Mechanical and Civil Engineering Department, 793003 Shillong, India*

Abstract. The reinforcement of ceramic materials with electrically conductive particles increases the overall conductivity of the ceramic material. This allows the ceramic material to be more readily machined using wire electrical discharge machining process. The current work is an approach to identify the machinability of multi walled carbon nanotubes filled alumina composites in wire electrical discharge machining process. Alumina samples of 5 vol. % and 10 vol. % multi walled carbon nanotubes are machined and analysed for material removal rate and the surface characteristics. An increase in material removal rate is observed with increase in filler concentrations. At the same time, better surface roughness is observed. The surface characteristics of composite alumina are further compared with Monel 400 alloy. It has been observed that spalling action is the dominating material removal mechanism for alumina composites, while melting and evaporation is for the Monel 400 alloy.

1 Introduction

Electrical Discharge Machining (EDM) and Wire Electrical Discharge Machining (WEDM) is a well-known non-conventional machining process for machining conductive materials. A wide range of materials can be successfully machined in this method by using a relatively softer tool material as long as they are conductive in nature. The machining phenomenon has been made successful as there is no relative contact between the tool and the work piece. In the normal machining scenario, the tool is made the cathode and the work piece material as anode. Both the tool and work piece are separated by a very small distance, known as the spark gap. The gap is then filled with a dielectric medium, generally hydrocarbon based oil in case of EDM, while deionised water for the case of WEDM. When current is supplied in the system, a potential difference is generated in the system. Due to the generated potential difference, the dissociation of the dielectric fluid occurs, which allows the transfer of electrons from cathode to anode. This leads to an intense generation of sparks from tool to the workpiece. The striking of sparks causes an intense localised heating that leads to melting of the work piece at that specific location. Here, most of the melted materials are either evaporated or flushed away under the action of the dielectric fluid. However, some portion of the work piece material gets re solidified and settles down on the surface as heat affected zone, commonly known as the recast layer.

A major challenge till date is the possibility to machine ceramic and its composites in EDM/WEDM process. It is a well-known fact that as the overall electrical conductivity of the ceramic material is very low. This restricts the usage of EDM/WEDM process to machine ceramic materials. However, researchers have identified two alternate approaches to machine ceramic in EDM/WEDM process. The first approach is the inclusion of a conductive layer, known as the Assisting Electrode (AE) on the surface of the electrically non-conductive ceramic [1]. The applied AE layer being conductive in nature helps in initial triggering of sparks. After the completion the AE layer, further machining is carried forward by the continuously generated pyrolytic carbon particles. The carbon particles are continuously generated due to the breaking and dissociation of kerosene (hydrocarbon based dielectric fluid). A continuous formation of the carbon layer on the surface of the work piece ensures further machining to propagate. Thus, dependence on hydrocarbon based dielectric oil is a must to sustain further machining. Various types of AE material were reported by various researchers that can successfully trigger the initial machining process [2-4]. Further, researchers have modified the general WEDM setup to a die sinking type WEDM to facilitate the use of kerosene [5-6]. A successful machining was reported in this regard. In the current approach, the overall material removal rate was found to be very low and the obtained surface finish was very rough.

Another approach that researchers adopted to machine ceramics in EDM/WEDM process is by reinforcing the ceramic material with conductive filler. This method increases the overall electrical conductivity of the ceramic material, thereby rending it to be machinable in EDM/WEDM [7-11]. Researchers have used TiN, TiC, TiCN, SiC, NbC, TiB_2 and WC as the reinforcing materials. An increase in the overall electrical conductivity of ceramic materials like Al_2O_3, ZrO_2 and Si_3N_4 were observed. This allowed machining of ceramic material with a much lesser ease. However, the percentage volume incorporation of such conducting fillers were found to be very high (approximately 30 vol. %). This alters the overall properties of the ceramic material as a whole. An improvement in this method is the use of carbon nanotubes (CNT) as the reinforcing conductive fillers. A small amount of the CNT particle can considerably increase the overall electrical conductivity of the composites [12]. Thus a good

[a] Corresponding author : dksarma@nitm.ac.in

 DOI: 10.1051/matecconf/201713601001

machining aspect has been obtained with CNT as the conductive filler. Further, as a strengthening means of this approach, Melk et al [13] reported that the use of multi walled carbon nanotubes (MWCNT) in zirconia ceramics at very low concentrations facilitate effective machining capabilities in EDM process. Singh et al [14] made an initial effort to identify the machining aspect of MWCNT filled alumina composites in WEDM process. Substantial machining capabilities were observed with 7.5 vol % MWCNTs or more. Keeping the above reported works in mind, the researchers have decided to further analyse the machinability aspect of MWCNT filled alumina composites. The variation of the material removal rate and the surface characteristics are identified and discussed in this paper. To get a better understanding of the surface characteristics, a comparison of the machining of alumina composites and Monel 400 alloy is represented.

2 Experimental Works

The main aim of this work is to identify the machinability aspect of MWCNT filled alumina composites in WEDM process and then to further compare it with Monel 400 alloy. Two samples of MWCNT concentration of 5 and 10 % by volume are considered [15]. It has been observed that the electrical conductivity is found to increase with the increase in CNTs, while the thermal conductivity decreases correspondingly (Table 1). The increase in electrical conductivity is put forward by the increasing concentrations of MWCNTs that increases the overall electrical network in the composites. On the other hand, decrease in thermal conductivity is attributed by the porosity of the composites and the grain boundary interactions. Anisotropy properties are observed in the samples due to random orientations of the CNTs. WEDM setup (DK7732, Concord Limited, Bangalore, India) has been used to perform the machining process (Fig. 1). The various input parameter used for the current machining process is represented in table 2. Further, the surface roughness of the machined samples are analysed using a surface roughness tester (Surtronic S-128, Taylor Hobson). A scanning electron microscope (SEM; JSM – 6360 JEOL) is then used to analyse the surface and the sub surface characteristics.

Table 1. Properties of MWCNT filled alumina composites

MWCNT	Density (kg/mm^3)	Electrical Conductivity (S/m)	Thermal Conductivity (W/m·K)
5 vol. %	3840.80	479	24.72
10 vol. %	3672.20	1402.50	15.38

Figure 1. WEDM setup.

Table 2. Input parametric conditions

Parameters	Values
Pulse on time (T_{on})	25 µs
Pulse off time (T_{off})	9 µs
Wire speed setting	0 (1493 rpm)
Current	2 A
Dielectric medium & conductivity	Deionised water & 0.7 µS/cm
Wire material & Diameter	Molybdenum & 0.18 mm

3 Results and Discussion

The experiments are performed on two types of MWCNT alumina composites and in Monel 400 alloy with the same input parametric conditions as shown in table 2. The experiments are performed with the material removal rate (MRR) and surface roughness (R_a) as the output performance characteristics. The experimental results are shown in table 3.

It has been found that the MRR increases considerably with increase in MWCNT concentration (Fig. 2). The increase in MRR has been brought forward by the incorporation of higher CNTs in the composites. With a comparatively higher CNTs in the composite, the electrical network increases, thereby increasing the overall passage of current. Thus, this condition ensures an increase in the net heat flux generated in between the tool and the alumina composites. Though the MRR is increasing for MWCNT alumina composite, it is not as prominent as that of Monel 400. The vast difference in MRR at a particular parametric setting is due to the immense difference in the electrical conductivity of the materials (Fig. 2). The Monel alloy being more conductive, have a higher tendency to allow the current to pass through. Thus, a higher material removal phenomenon is observed.

Table 3. Experimental results

Material Types	Material removal rate (mm^3/min)	Surface Roughness (μm)
5 vol. % MWCNT alumina composite	1.91	3.167
10 vol. % MWCNT alumina composite	9.69	2.28
Monel 400 alloy	16.27	3.56

During machining of 5 vol. % MWCNT filled alumina composites, some wire lag phenomena is observed. The condition arises as a consequence of the low conductivity of the material that restricts the effective machining capabilities. In some case, the sample breaks off before actual completion of the machining process due to wire lag phenomena. The wire lag is also expected during machining of 10 vol. % MWCNT filled alumina composites. But, no appreciable wire lag was observed. This may be due to the overall increase in the electrical conductivity of the alumina composites with increase in MWCNT concentrations. On the other hand, wire lag phenomena is not observed in machining of Monel 400 alloy. This is due to the fact that Monel 400 alloy being highly conductive in nature; efficient machining capabilities can be obtained.

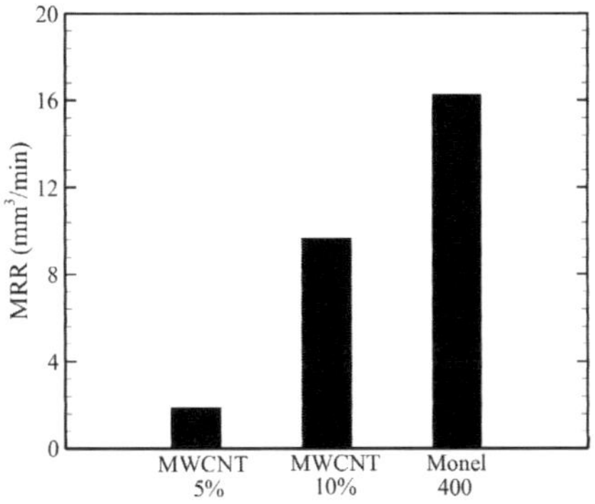

Figure 2. Variation of MRR of 5, 10 vol. % MWCNTs alumina composites and Monel 400

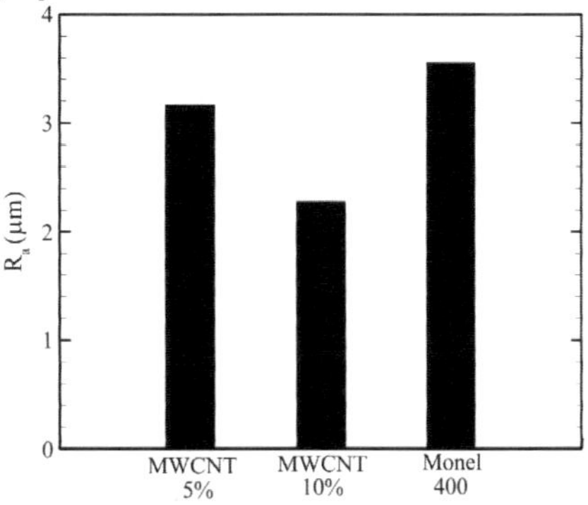

Figure 3. Variation of R_a of 5, 10 vol. % MWCNTs alumina composites and Monel 400

Figure 3 shows the plot of surface roughness of the alumina composites and Monel 400 alloy. As expected, better R_a is observed with increase in the MWCNT concentrations. During machining of conductive ceramic composites, the passage of electrical network is mainly due to the conductive reinforced fillers. Hence, the heating of reinforced fillers occurs initially. The intense generated heat is then dissipated to the neighbouring ceramic matrix materials. This leads to melting of the ceramic material by spalling action. Since the CNT concentration of 5 vol. % MWCNT alumina composites is less, so the rate of heat transfer from the CNT to the alumina material is found to be lesser. Hence, the associated R_a is found to be quite rough, coupled with a low MRR. However, for the case of 10 vol. % MWCNT alumina composites, more alumina materials get heated due to the increasing CNT concentrations. Hence, a lower R_a is obtained, coupled with a higher MRR. For the case of Monel 400 alloy, an intense localised heating occurs due to high electrical conductivity. At the same time, due to the inability to remove all the melted materials by the action of the cooling nature of the dielectric fluid, a portion of the melted material gets re solidified on the surface and forms a recast layer. The recast layer formation is generally associated with a high debris deposition on the surface. Therefore, a comparatively high R_a value is observed in Monel 400.

Figure 4 shows the SEM images to compare the surface characteristics of 5 vol. %, 10 vol. % MWCNT alumina composite along with Monel 400 alloy. A mixed spalling and melting phenomena is observed as the basic material removal characteristic in case of 5 and 10 vol. % MWCNT alumina composites. However, for Monel 400 alloy, the melting and evaporation is seen as the material removal characteristics. For 5 vol. % MWCNT alumina composite, the surface is highly porous in nature with ample amount of recast layer (Fig. 4a). However, for 10 vol. % MWCNT alumina composite, the machined surface is found less porous in nature (Fig. 4b). The irregular debris particles on the machined surface are due to the effect of spalling mechanism while regular spherical debris due to the melting and evaporation effect. Thus, it is observed that spalling effect is more prominent with 10 vol. % MWCNT alumina composites, while melting and evaporation is the dominant factor for Monel 400(Fig. 4c).

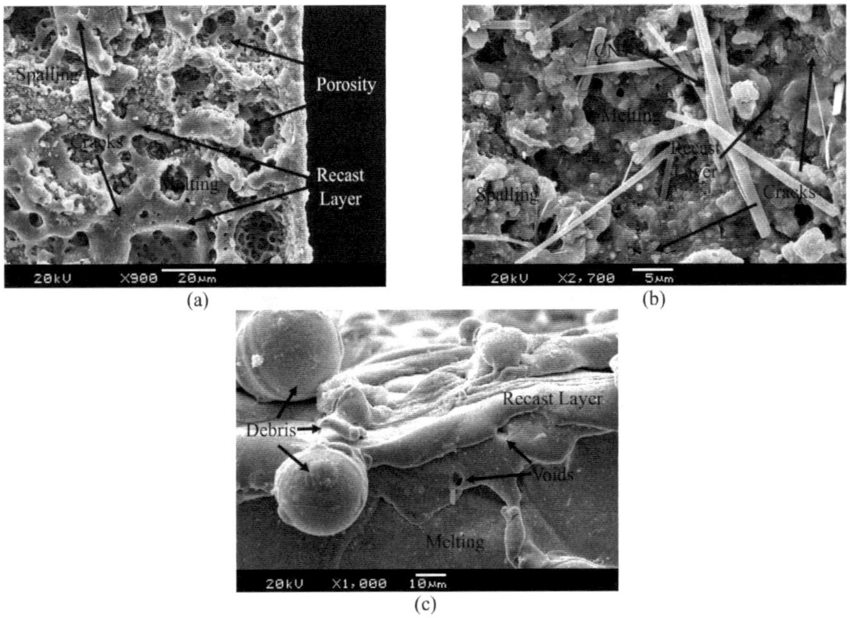

Figure 4. SEM images for surface characteristic analysis of (a) 5 vol. %, (b) 10 vol. % MWCNT filled alumina composites and (c) Monel 400 alloy.

The fabricated MWCNT alumina composites are found to be brittle in nature [15]. So, the probability of cracks formation is quite high. Figure 5 represents the SEM images of cracks formation and propagation during machining of 5 and 10 vol. % MWCNT alumina composites. As alumina composites with 5 vol. % MWCNTs being more brittle and of low electrical conductivity, longer span of micro cracks are observed. While, a comparatively shorter span of micro cracks are observed in samples with 10 vol. % MWCNTs due to higher electrical conductivity. The appearance and formation of cracks disrupts the surface characteristics. Also, it has a more probability of breakage of the work piece, which is observed during machining of 5 vol. % MWCNT filled alumina composites. Presence of sub surface micro cracks after the recast layers are also found on a closer analysis of the machined sample (Fig. 6). This shows the high susceptibility of the material under the influence of rapid thermal expansion. Further, it is expected that the high brittleness of the material have contributed to the surface micro cracks and sub surface cracks formation.

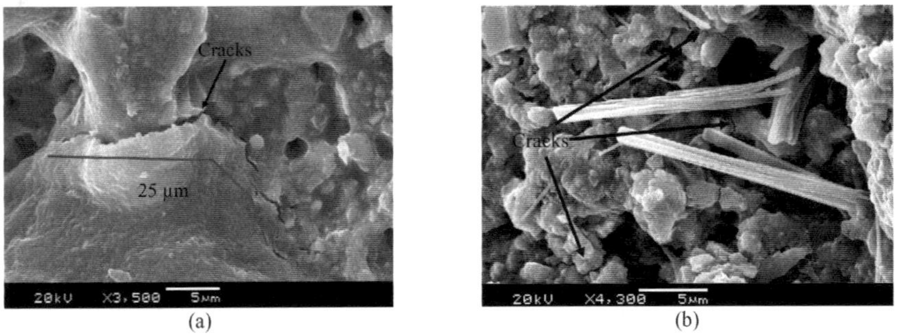

Figure 5. SEM micrograph for formation and propagation of cracks in (a) 5 vol. % and (b) 10 vol. % MWCNT filled alumina composites.

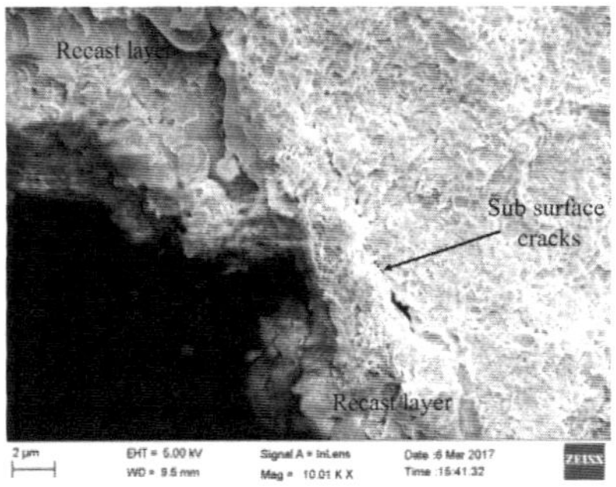

Figure 6. Sub surface analysis of 10 vol. % MWCNT filled alumina composites.

 DOI: 10.1051/matecconf/201713601001

4 Conclusions

A WEDM process has been performed to machine MWCNT filled alumina composites. Two concentrations viz., 5 and 10 vol. % MWCNTs are reinforced as the conducting phase in alumina. A more stable machining phenomenon is obtained with 10 vol. % MWCNT filled alumina composites. The machining process of 5 vol. % MWCNT filled alumina composites was found to be difficult owing to its low electrical conductivity. Some wire lag phenomena was also observed during the course of machining. The tendency of formation and occurrence of cracks are found to be more prominent at lower CNT concentration. A comparison of the alumina composites has been made with that of Monel 400 alloy with regards to machinability characteristics. Unlike metals where melting and evaporation is the main material removal mechanism, spalling effect was found to be more dominating for the case of ceramic composites.

References

1. N. Mohri, Y. Fukuzawa, T. Tani, N. Saito, K. Furutani, CIRP Ann – Manuf Techn, 45(1) (1996)
2. Y. Guo, G. Dang, J. Bai, Z. Lu, Key Eng Mater, 375-376 (2008)
3. N. Mohri, Y. Fukuzawa, T. Tani, T. Sata, CIRP Ann – Manuf Techn, 51(1) (2002)
4. A. Sabur, M. Y. Ali, M. A. Maleque, A. A. Khan, Procedia Eng, 56 (2013)
5. J Kozak, K. P. Rajurkar, N. Chandarana, J Mater Process Tech, 149 (2004)
6. Y. Guo, P. Hou, D. Shao, Z. Li, L. Wang, L. Tang, Mater and Manuf Process, 29 (2014)
7. C. Martin, B. Cales, P. Vivier, P. Mathieu, Mat Sci Eng A – Struct, 109 (1989)
8. T. Matsuo, E. Oshima, CIRP Ann – Manuf Techn, 41(1) (1992)
9. E. Ferraris, D. Reynaerts, B. Lauwers, CIRP Ann – Manuf Techn, 60 (2011)
10. R. Landfried, F. Kern, R. Gadow, Int J Appl Ceram Tec, 10(3) (2013)
11. K. M. Patel, P. M. Pandey, P. V. Rao, Int J Refract Met H, 27(5) (2009)
12. O. Malek, J. G. Julian, J. Vleugelss, W. Vanderauwera, B. Lauwers, M. Belmonte, Mater Today, 14(16) (2011)
13. L. Melk, M. L. Antti, M. Anglada, Ceram Int, 42(5) (2016)
14. M. A. Singh, D. K. Sarma, O. Hanzel, J. Sedlacek, P. Sajgalik, J Eur Ceram Soc, (2017)
15. O. Hanzel, J. Sedlacek, P. Sajgalik, J Eur Ceram Soc, 34 (2014)

A few simple CFD models for flow simulation around groyne

Hriday Mani Kalita[1,a], Arup Kumar Sarma[2], Comingstarful Marthong[1] and Deba Kumar Sarma[3]

[1]*Civil Engineering Department, National Institute of Technology Meghalaya, Meghalaya, India, Pin: 793003*
[2]*Civil Engineering Department, Indian Institute of Technology Guwahati, Assam, India, Pin: 781039*
[3]*Mechanical Engineering Department, National Institute of Technology Meghalaya, Meghalaya, India, Pin: 793003*

Abstract. The flow processes around a groyne structure is a complex and challenging topic in computational hydraulics and has brought attention of many researchers in the last few decades. In comparison to experimental studies, numerical modeling can provide more details about the flow characteristics at less cost. In the present work, performances of four simple finite difference schemes, i.e., Lax diffusive scheme, predictor corrector scheme, modified predictor corrector scheme and Beam and Warming scheme are compared in simulation of flow around groyne structure. The numerical results are validated with available experimental data and the best model is found out in terms of accuracy and efficiency.

1 Introduction

Any kind of structures those are constructed in a river for guiding/managing the river water for beneficial purpose of mankind are termed as river training structures. Groyne or spur dike is one of the most commonly used river training structures generally constructed from the river bank and extend towards the main channel flow. They are constructed for various purposes including, deflecting/repelling the flow away from bank for safety, increasing flow depth for smooth navigation, etc. Construction of these structures leads to blockage in the flow and always leads to changes in the local hydrodynamics as well as the morphology of the river. At the same time, the flowing water also imparts some effects on the groyne. These effects may be in the form of increment in the flow speed near the tip of the structure, development of local scouring around the structure due to pressure difference between the top and bottom water pressures near the structure. These local scouring many a time even may leads to failure of the structure itself. Moreover, the opposite river bank also faces high speed and shear stress upon construction of these structures. Therefore, before implementing a project deals with construction of groyne, it is always essential to conduct a model study to evaluate the probable effect of the groyne on the river as well as the reverse effect of the river on the structure itself. These model studies however, may be based on either experimental approaches or numerical approaches. In experimental studies, the whole groyne-river system is physically developed in some appropriate scales and experiments are conducted over this system. Ahmad [1] conducted one of the first experimental studies for simulating scouring process around groynes. In his study he observed that the water level remains constant upstream of the groyne, followed by a sudden drop near the tip and then gradually recovers along the downstream directions. However, he did not measure the flow velocity near the vicinity of the groyne. Rajaratnam and Nwachukwu [2] were the first researchers

[a] Corresponding author : hridaymanikalita@gmail.com

who tried to measure the flow velocity around a groyne. They conducted different set of experiments by changing the shape of groyne, flow depth, etc., and observed that groyne causes significant disturbances to the flow for a short distance upstream of it and for a longer distance downstream of it. From that time onward, several experimental studies were carried out by numerous researchers [3-4] for experimental simulation of groyne. However, these experimental models are subjected to some shortcomings like, high investment cost, scale issues, time consuming and limited measuring points. It has been observed that numerical models have the capacity to provide more details than physical models at low cost. Regarding the numerical simulation, the shallow water equations [5] can reasonably represent the flow characteristics around a groyne. In the recent decades, due to advancement in the computer power, researchers have also started using numerical models in this regard. Tingsanchali and Maheswaran [5] developed one of the first numerical models to simulate the flow of water around a groyne by solving the governing shallow water equations using a finite difference method. After that several two dimensional (2D) and three dimensional (3D) numerical models [6, 7] have been proposed by many investigators to replicate the complex flow processes around groyne. However, the flow processes around a groyne is very complex and therefore before using the numerical results for practical purposes a detail validation of these results with experimental data is still required. From the literature it appears that even though a large number of numerical models are available for flow simulation around groyne very less attention was given for comparing performances of different models. In this work four simple finite difference numerical models are presented for comparing and assessing the performances of the respective models on the basis of available flume data.

2 Governing equations and numerical techniques

2.1 Governing equations

The governing shallow water equations can be derived by vertically integrating the Navier Stokes equations over the flow depth under the assumptions of hydrostatic pressure distribution and small bottom slope. The governing equations are transformed from the physical coordinate system (x, y) to a boundary fitted coordinate system (ξ, η) to make the numerical model compatible with complex boundaries and can be written as [8];

$$P_t + Q_\xi + R_\eta + S = 0 \qquad (1)$$

where

$$P = J\begin{Bmatrix} h \\ hu \\ hv \end{Bmatrix}, Q = J\begin{Bmatrix} hU \\ huU + \xi_x\left(\frac{gh^2}{2}\right) \\ hvU + \xi_y\left(\frac{gh^2}{2}\right) \end{Bmatrix}, R = J\begin{Bmatrix} hV \\ huV + \eta_x\left(\frac{gh^2}{2}\right) \\ hvV + \eta_y\left(\frac{gh^2}{2}\right) \end{Bmatrix}, S = J\begin{Bmatrix} 0 \\ -gh\left(S_x - \frac{n^2 \times u \times \sqrt{u^2+v^2}}{h^{\frac{4}{3}}}\right) \\ -gh\left(S_y - \frac{n^2 \times v \times \sqrt{u^2+v^2}}{h^{\frac{4}{3}}}\right) \end{Bmatrix}$$

h is depth of flow, *u*, *v*, *U*, *V* are velocity components in *x*, *y*, *ξ* and *η* directions, respectively, g is the acceleration due to gravity, S_x and S_y are bed slopes is x and *y* direction, respectively, n is Manning's roughness coefficient. The jacobian *J* and the metrices (ξ_x , ξ_y , η_x , η_y) are results of grid transformation and are calculated using central finite difference approximations [8].

2.2 Numerical solution of the governing equations

The governing equations presented by equation (1) constitute a system of mixed hyperbolic partial differential equations and are difficult to solve analytically without considering large number of approximations. Therefore, these equations are mostly solved using numerical techniques. In this work four different numerical techniques based on finite difference approximations are used in this regard and are presented in the subsequent sections.

2.2.1 Lax diffusive scheme

Lax diffusive scheme is a finite difference scheme based on explicit approach and is first order in time and second order accurate in space. The major advantages of this scheme are, single step in nature leading to less computational time requirement and no need of artificial damping/viscosity. Applying finite difference formulation of Lax diffusive scheme for the equation (1) yields [9];

$$P_{i,j}^{k+1} = \frac{1}{4}\left(P_{i,j+1}^{k} + P_{i,j-1}^{k} + P_{i+1,j}^{k} + P_{i-1,j}^{k}\right) - \frac{\Delta t}{2\Delta\xi}\left(Q_{i,j+1}^{k} - Q_{i,j-1}^{k}\right) - \frac{\Delta t}{2\Delta\eta}\left(R_{i+1,j}^{k} - R_{i-1,j}^{k}\right) - \frac{\Delta t}{4}\left(S_{i,j+1}^{k} + S_{i,j-1}^{k} + S_{i+1,j}^{k} + S_{i-1,j}^{k}\right) \quad (2)$$

where, $\Delta\xi$ is the grid spacing in flow direction designated by the subscript *j*, $\Delta\eta$ is the grid spacing in transverse direction designated by the subscript *i*, Δt is the grid spacing in the time axis (time step) designated by the superscript *k*.

2.2.2 Mac-Cormack predictor corrector scheme

Mac-Cormack predictor corrector scheme is another explicit finite difference scheme which is second order accurate in both time and space. This technique is composed of two steps called predictor and corrector steps. Using backward difference in the predictor step an initial solution is obtained which is further corrected in the corrector step using forward difference. Final solution is obtained by taking average of these predicted and corrected solutions. Some of the advantages associated with this scheme are, capacity to handle mixed sub and super critical flow and efficient simulation of bore or shock waves. The governing equations can be discretized with this scheme as [9];

$$P_{i,j}^{p} = P_{i,j}^{k} - \frac{\Delta t}{\Delta\xi}\left(Q_{i,j}^{k} - Q_{i,j-1}^{k}\right) - \frac{\Delta t}{\Delta\eta}\left(R_{i,j}^{k} - R_{i-1,j}^{k}\right) - \Delta t S_{i,j}^{k} \quad (3)$$

$$P_{i,j}^{c} = P_{i,j}^{k} - \frac{\Delta t}{\Delta\xi}\left(Q_{i,j+1}^{k} - Q_{i,j}^{k}\right) - \frac{\Delta t}{\Delta\eta}\left(R_{i+1,j}^{k} - R_{i,j}^{k}\right) - \Delta t S_{i,j}^{p} \quad (4)$$

$$P_{i,j}^{k+1} = \frac{1}{2}\left(P_{i,j}^{p} + P_{i,j}^{c}\right) \quad (5)$$

2.2.3 Modified predictor corrector scheme

The governing equations presented above have an inherent property of signal propagation i.e., information comes only from the upstream direction in case of super critical flow and information comes from both upstream and downstream in case of subcritical flow. Therefore, in this modified version of the predictor corrector scheme backward finite difference is used in predictor step and forward finite difference is used in corrector step for subcritical region. However, for supercritical region, both steps employ backward difference to eliminate erroneous influence from the downstream flux on the computed values. This simple technique has made this scheme efficient in handing mixed flow regime. The governing equations can be discretized as follows;

$$P_{i,j}^{p} = P_{i,j}^{k} - \frac{\Delta t}{\Delta \xi}\left(Q_{i,j}^{k} - Q_{i,j-1}^{k}\right) - \frac{\Delta t}{\Delta \eta}\left(R_{i,j}^{k} - R_{i-1,j}^{k}\right) - \Delta t S_{i,j}^{k} \tag{6}$$

a) If $\sqrt{u^2 + v^2} < \sqrt{gh}$, subcritical flow region,

$$P_{i,j}^{c} = P_{i,j}^{k} - \frac{\Delta t}{\Delta \xi}\left(Q_{i,j+1}^{k} - Q_{i,j}^{k}\right) - \frac{\Delta t}{\Delta \eta}\left(R_{i+1,j}^{k} - R_{i,j}^{k}\right) - \Delta t S_{i,j}^{p} \tag{7}$$

b) If $\sqrt{u^2 + v^2} > \sqrt{gh}$, supercritical flow region,

$$P_{i,j}^{c} = P_{i,j}^{p} \tag{8}$$

2.2.4 Beam and Warming scheme

Beam and Warming scheme is an implicit finite difference scheme and second order accurate in both time and space. The major advantages of this scheme are its non-iterative nature and quick converging capacity. Here, the solution advances in a two-step sequence and each step requires solution of a block-tridiagonal system of equations. In the present work Thomas algorithm is used for efficient solution of the block-tridiagonal systems. The equations solved in ξ and η directions and the final values at the next time level can be computed as [9];

$$\left(I + \frac{2\Delta t}{3}\frac{\partial}{\partial \xi}A^{k}\right)P^{*} = -\Delta t \frac{3}{2}\left(\frac{\partial Q}{\partial \xi} + \frac{\partial R}{\partial \eta} + S\right)^{k} + \frac{1}{3}\left(P^{k} - P^{k-1}\right) \tag{9}$$

$$\left\{I + \frac{2\Delta t}{3}\left(\frac{\partial}{\partial \eta}B + C\right)^{k}\right\}P^{**} = P^{*} \tag{10}$$

$$P_{i,j}^{k+1} = P_{i,j}^{k} + P_{i,j}^{**} \tag{11}$$

2.2.5 Stability and Artificial viscosity

The stability of the present models is represented using the Courant-Friedrichs-Lewy (CFL) criteria which restricts the time step value as;

$$\Delta t = \frac{Cr \times \min \times \min\left(\Delta x, \Delta y\right)}{\max\left[\left(u + \sqrt{gh}\right), \left(v + \sqrt{gh}\right)\right]} \tag{12}$$

where *Cr* is the Courant number and its value is limited to one for explicit schemes.

Another issue that arises while using second order accurate schemes is the dispersion errors and can be effectively removed by using artificial viscosity [9]. In the present work, the standard and the modified predictor corrector methods use a technique proposed by Jameson et al. [10] and the Beam and Warming methods uses a technique proposed by Klonidis and Soulis [11].

3 Application of the models for flow simulation around groyne

In order to appraise the performances of the present numerical schemes these are used to replicate the experimental study done by Rajaratnam and Nwachukwu for flow around groyne. In their study the flow channel was straight and rectangular in shape. The groyne considered was an aluminium plate of length 0.15 m. The numerical models have been developed for an area extending from 1.8 m upstream to 3.6 m downstream of the groyne. A finite difference grid of size 97×45 is created for the flow domain. Manning's roughness value for the channel was 0.01. An upstream flow of 0.045 cumec and a downstream depth of 0.189 m were used as boundary conditions. The channel bank and the groyne are simulated as reflective boundary technique. The *Cr* value used for the explicit schemes is 0.83 and the same for the implicit scheme is 2.2. Figures 1a, 1b, 2a and 2b show the velocity vector plots for the Lax diffusive, predictor corrector, modified predictor corrector and Beam and Warming schemes, respectively. It appears from the results that the diffusive scheme produces a very small recirculation zone downstream of the groyne. It may be due to excessive diffusion associated with the scheme. However, a clear recirculation zone ending at 3.6 m is observed for the rest schemes exactly matching with the experiment.

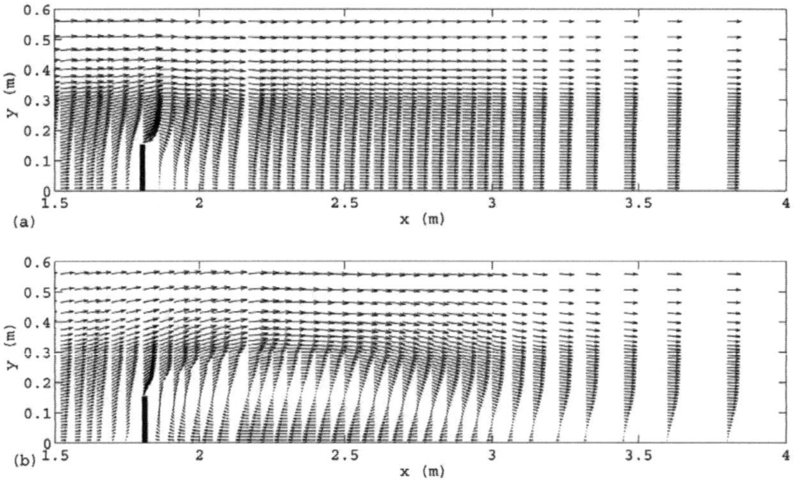

Figure 1 Velocity vector plot: a) Lax diffusive scheme, b) Predictor corrector scheme

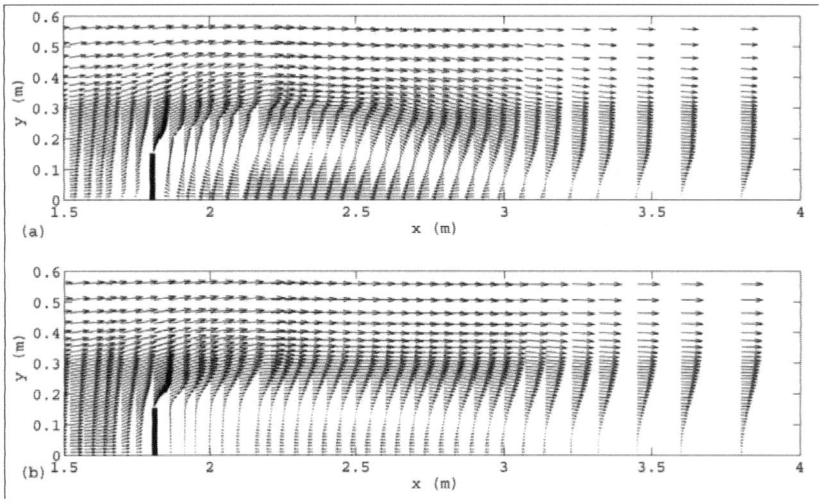

Figure 2 Velocity vector plot: a) Modified predictor corrector scheme, b) Beam and Warming scheme

DOI: 10.1051/matecconf/201713601002

Apart from the recirculation zone comparison, the resultant velocity is also compared with the experimental data. For this purpose, the resultant velocity (W) is first non-dimensionalized with respect to the upstream velocity of 0.253 m/s. Figures 3a, 3b, 3c and 3d show the comparison of simulated flow speed and experimental flow speed at four horizontal lines at distances of 0.15, 0.225, 0.4 and 0.6 m from the southern bank, respectively. From these results it can be concluded here that the Beam and Warming scheme has the good potential for accurate simulation of flow characteristics around groyne.

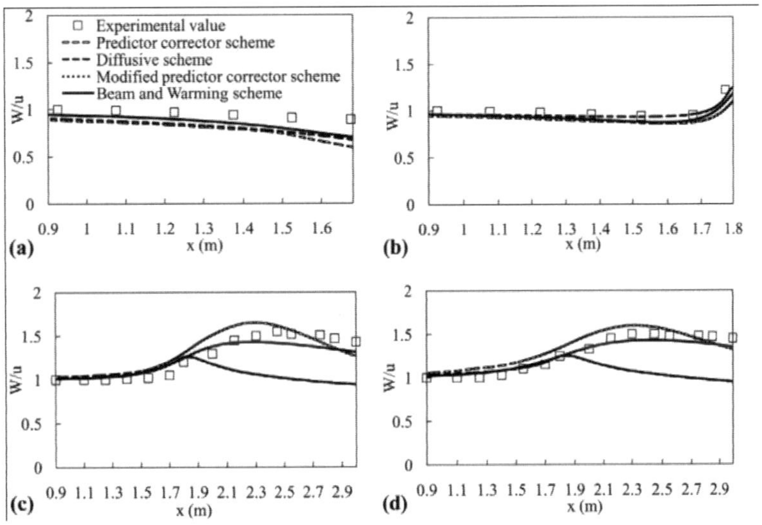

Figure 3 Comparison of flow speed: a) y=0.15 m, b) 0.225 m, c) 0.4 m, d) 0.6 m

4 Conclusions

In this study, four simple finite difference based models, i.e., Lax diffusive, predictor corrector, modified predictor corrector and Beam and Warming, are developed to simulate the flow characteristics around groyne within flowing water. In order to check the capability of the developed models, the results are compared with an available experimental data. It appears from the results that the Lax scheme is not suitable for the affirmed task as it contains excessive diffusion. Out of the rest three schemes, Beam and Warming scheme is found to very much accurate in flow simulation around groyne. Moreover, this scheme also remains stable for higher values of temporal steps.

References

1. M. Ahmad, *Minnesota Int. Hydraul. Convention* (1953)
2. N. Rajaratnam, B. Nwachukwu, J. Hydraul. Eng. **109** (1983)
3. R.A. Kuhnle, C.V. Alonso, F.D. Shields, J. Hydraul. Eng. **128** (2002)
4. A. Safarzadeh, S.A.A. Salehi, A.R. Zarrati, J. Hydraul. Eng. **142** (2016)
5. T. Tingsanchali, S. Maheswaran, J. Hydraul. Eng. **116** (1990)
6. A. Molinas, Y.I. Hafez, J. fluids structures **14** (2000)
7. J. Yazdi, H. Sarkardeh, H.Md. Azamathulla, A.Ab. Ghani, Int. J. River Basin Manag. **8** (2010)
8. D.A. Anderson, J.D. Tannehill, R.H. Pletcher, *Computational Fluid Mechanics and Heat Transfer* (McGraw-Hill, New York, 1984)
9. M.H. Chaudhry, *Open channel flow* (Prentice-Hall Inc, Englewood Cliffs, 2008)
10. A. Jameson, W. Schmidt, E. Turkel, *AIAA 14th Fluid And Plasma Dynamics Conference* (1981)
11. A.J. Klonidis, J.V. Soulis, J. Hydraul Res. **39** (2001)

Microstructure of Semi-Solid 6063 Alloy Fabricated by Radial Forging Combined with Unidirectional Compression Recrystallization and Partial Melting Process

Yongfei Wang[1,a], Shengdun Zhao[1] and Xuzhe Zhao[2]
[1]*School of Mechanical Engineering, Xi'an Jiaotong University, Xi'an 710049, China*
[2]*Purdue Institute of Technology, Purdue University, West Lafayette 47906, USA*

Abstract. Radial forging combined with unidirectional compression (RFCUM) is introduced in recrystallization and partial melting (RAP) to fabricate semi-solid 6063 aluminum alloy, which can be defined as a process of RFCUM-RAP. In this study, the microstructures of semi-solid 6063 alloy prepared by semi-solid isothermal treatment (SSIT) and RFCUM-RAP processes are investigated. The results show that, the solid grains of semi-solid alloy prepared by SSIT are large and irregular. However, solid grains of semi-solid billet prepared by RFCUC-RAP are fine and spherical. Additionally, during RFCUC-RAP process, with the increase of isothermal holding time, the shape of solid grain is more and more spherical, but the size of solid grain is gradually increased. To obtain ideal semi-solid microstructure, the optimal isothermal holding temperature and time are 630 °C and 5~10 min, respectively.

1 Introduction

Semi-solid metal process (SSMP) developed by Spencer and co-workers has been widely studied in recent years, which is an effective near-net-shape forming process [1-4]. Compared with conventional casting and forging processes, it offers significant advantages, such as easy forming, long die life, good microstructure and good mechanical properties [5]. The key requirement of SSMP process is that the semi-solid billet with fine and spherical solid grains uniformly distributed in the liquid phase. To prepare ideal semi-solid microstructures, many methods have been developed, such as, mechanical and electromagnetic stirring [6, 7], cooling slope [8], mechanical vibration [9], strain-induced melt activation (SIMA) [2, 10], and recrystallization and partial melting (RAP) [11]. Among these methods, RAP process consists of warm deformation and semi-solid isothermal treatment (SSIT) is a promising method, because that many alloys are always supplied in the deformed state.

Al-Mg-Si based aluminum alloys, such as 6063 alloy, are widely employed in automotive and aerospace industries due to their high mechanical properties. However, they are currently machined from the wrought state with much waste. Therefore, there is a strong driving force to use SSMP process to form this alloy. In this study, RFCUM is introduced in RAP to fabricate semi-solid 6063 alloy, the microstructures of semi-solid 6063 alloy prepared by semi-solid isothermal treatment (SSIT) and RFCUM-RAP process are investigated. Moreover, the optimal SSIT parameters for ideal semi-solid microstructures are also explored.

[a] Corresponding author : yongfeio@126.com

2 Experimental procedure

The as-extruded 6063 wrought aluminum alloy bar (⌀ 120 mm×180 mm) was used as experimental material, and it consisted of 0.93 wt.% Mg, 0.61 wt.% Si, 0.26 wt.% Fe, 0.19 wt.% Cu, 0.15 wt.% Mn, 0.05 wt.% Ti, 0.11 wt.% Cr and balance of Al. The solidus and liquidus temperatures were 615 °C and 655 °C, respectively, which were determined by differential scanning calorimeter (DSC) method. Moreover, the solid fraction versus temperature curve obtained from DSC result is shown in Figure. 1.

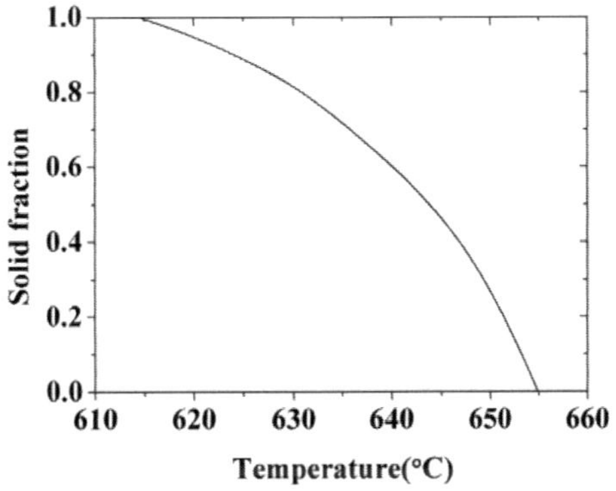

Figure 1. Liquid fraction versus temperature curve derived from DSC result.

Figure. 2 shows the schematic illustration of RFCUC-RAP process. As shown in Figure. 2, the RFCUC-RAP process could be divided into two steps: the deformation and the SSIT steps. The deformation process of 6063 alloy consists of radial forging and unidirectional compression. First, the as-received 6063 bar after preheating at 300 °C for 90 min was radial forged with 70% ARR and quenched in water. The ARR can be defined as area reduction ratio, which is calculated by the following equation: ARR=(A0-A1)/A0, where A0 and A1 are the initial and final cross sectional area of the alloy bar, respectively. Second, the as-forged material with a diameter of 10 mm and a length of 15 mm was cut along the radial forging direction for unidirectional compression. The compression ratio was 80%, and the compression direction was the same with the radial forging direction. During the SSIT process, the samples were treated at 630 °C for 1~20 min in a resistance furnace. After the SSIT process, the samples were taken out immediately and quenched in water to obtain the semi-solid microstructures. In addition, in order to investigate the effects of SSIT process on the microstructure of 6063 alloy, the samples cut from the as-received alloy were treated at 630 °C for 5~20 min, and then quenched in water.

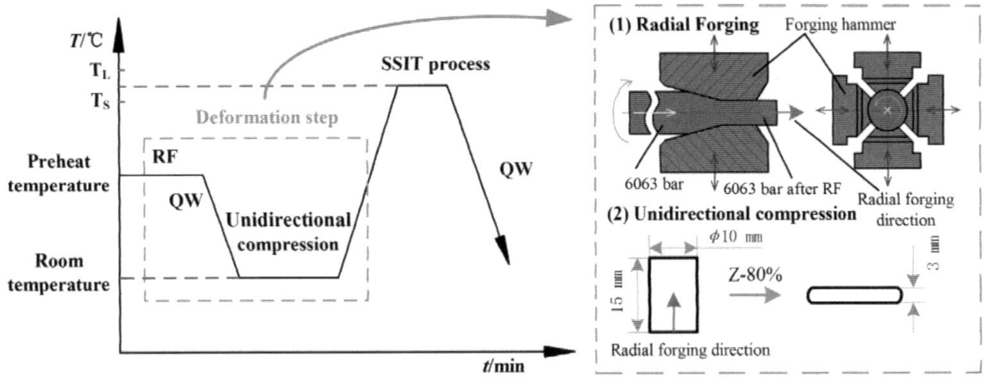

Figure 2. Schematic illustration of RFCUMRAP process.

The samples after SSIT process of as-received and RFCUC deformed alloys were ground and polished, and then etched with an aqueous solution of 5% HF. The metallographic observation was carried out by optical microscopy (OM). The average grain size (D) and shape factor (F) of solid grains were measured by image analysis software (Image-Pro Plus 6.0), and calculated in each case by Eqs. (1) and (2) [12]:

$$D = \frac{\sum_{N=1}^{N} \sqrt{4A/\pi}}{N} \tag{1}$$

$$F = \frac{\sum_{N=1}^{N} 4\pi A / P^2}{N} \tag{2}$$

Where A, N and P are area, the number and perimeter of solid grains, respectively. If F is gradually closed to 1, solid grains will be more and more spherical.

3 Results and discussion

3.1 Microstructure of as-received and RFCUC-deformed alloy

Figure. 3 shows the microstructure of as-received and RFCUC-deformed alloy. As shown in Figure. 3(A), the microstructure is typical of extrusion alloy with fibrous grains and intermetallic particles aligned in the extrusion direction. The microstructural feature of RFCUC deformed alloy shown in Figure. 3(B) is similar to that of the as-received alloy. However, it is evident that the deformation degree shown in Figure. 3(B) is higher than that in Figure. 3(A), resulting from the additive deformation of RFCUC process. Additionally, there is no evidence of recrystallization across can be seen in Figure. 3(B), possibly owing to the low deformation temperature [13].

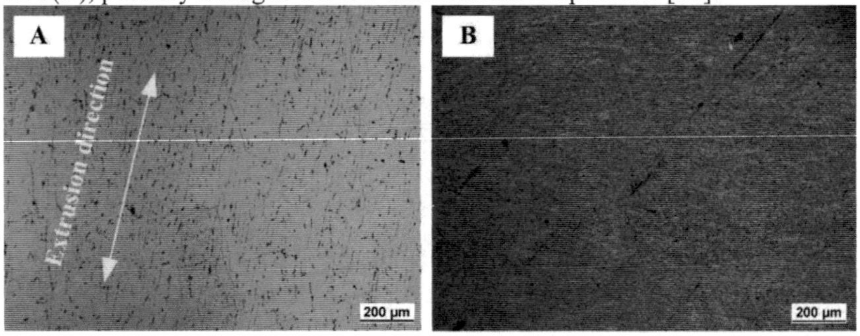

Figure 3. Microstructure of 6063 alloy: (A) as-received alloy, (B) RFCUC-deformed alloy.

3.2 Microstructure of as-received alloy after SSIT

Figure. 4 shows the microstructure of as-received alloy after SSIT at 630 °C for 5~20 min. In Figure. 4(A), when the isothermal holding time is 5 min, the liquid phase is little; the number of independent solid grain is few, and the size of solid grain is large. In Figure. 4(B), when the isothermal holding time is increased to 10 min, the number of solid grains is increased, and the size of solid grain is decreased, indicating that the size of solid grain can be decreased at the intial stage of the SSIT process. With increasing the isothermal holding time from 15 min to 20 min, the solid grains (Figure. 4(C) and (D)) become larger and more irregular than that shown in Figure. 4(A) and (B). The main reason can be explained as follows: the as-received 6063 alloy is an extrusion bar with little strain energy, which can not offer an enough driving force for recovery and recrystallization during the SSIT process [14]. The optimal average grain size and shape factor of solid grains measured from Figure. 4 are about 207 μm and 0.72 (Figure. 4(B)), respectively. For the SSMP process, the semi-solid billet with large and irregular solid grain will have an adverse effect on the mechanical properties of final parts [15]. Therefore, it is obvious that the SSIT of as-received alloy is unsuitable for fabricating ideal semi-solid 6063 alloy.

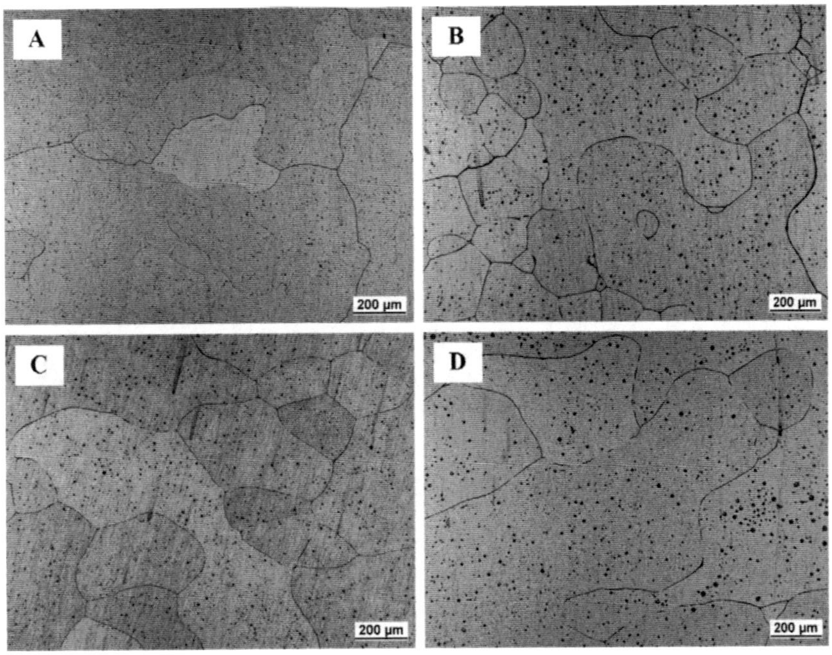

Figure 4. Microstructure of as-received alloy after SSIT at 630 °C for different times: (A) 5 min, (B) 10 min, (C) 15 min and (D) 20 min.

3.3 Microstructure of RFCUC-deformed alloy after SSIT

Figure. 5 shows the microstructure of RFCUC-deformed alloy after SSIT at 630 °C for 1~20 min. As shown in Figure. 5 (A), when the isothermal holding time is 1 min, the deformed direction of alloy are gradually disappeared, and some recrystallized grains can be found in the microstructure (marked with A1). As indicated in Figure. 5(B), when the isothermal holding time is 3 min, the recrystallization phenomenon fully occurs. At the same time, the boundaries of solid grains start to melt. The liquid phase starts to appear at the grain boundaries and triple points between grains (marked with B1). As shown in Figure. 5(C), when the isothermal holding time is increased to 5 min, the liquid is gradually increased because of the melting process; the majority of solid grains are separated by liquid film, and the grain size is small. As shown in Figure. 5 (D)~(F), when the isothermal holding time is prolonged from 10 min to 20 min, the liquid film is more and more thicker due to the increase of liquid phase, the grain size is gradually increased and the spheroidization degree of grain is improved. Moreover, some excessive larger solid grains (marked with F1 and F2) can be found in Figure. 5(F), which is attributed to the grain coalescence.

Figure. 6 shows the variations of the average grain size and shape factor of solid grains with the isothermal holding time from the quantitative analysis of the microstructure shown in Figure. 5. It should be noted that it is difficult to measure the solid grains shown in Figure. 5(A) and (B), because there are no obvious solid grains shown in Figure. 5(A) and the solid grains shown in Figure. 5(B) are connected together. As shown in Figure. 6, with the increase of isothermal holding time, the average grain size and shape factor of solid grains are gradually increased.

The microstructural evolution of alloy produced by the RFCUC-RAP process can be described as follows: during the RFCUC process of alloy, the strain energy is stored in the forms of dislocations, providing the driving force for recovery and recrystallization [16]. Therefore, during the initial stage of SSIT process, in order to reduce the free energy, the dislocation will climb and cross slip, leading to the occurrence of recovery and recrystallization (Figure. 5(A)). With the increase of isothermal holding time, liquid phase starts to appear (marked with B1) because that the isothermal holding temperature (630 °C) is higher than the solidus temperature. Moreover, the liquid penetrates the grain boundaries

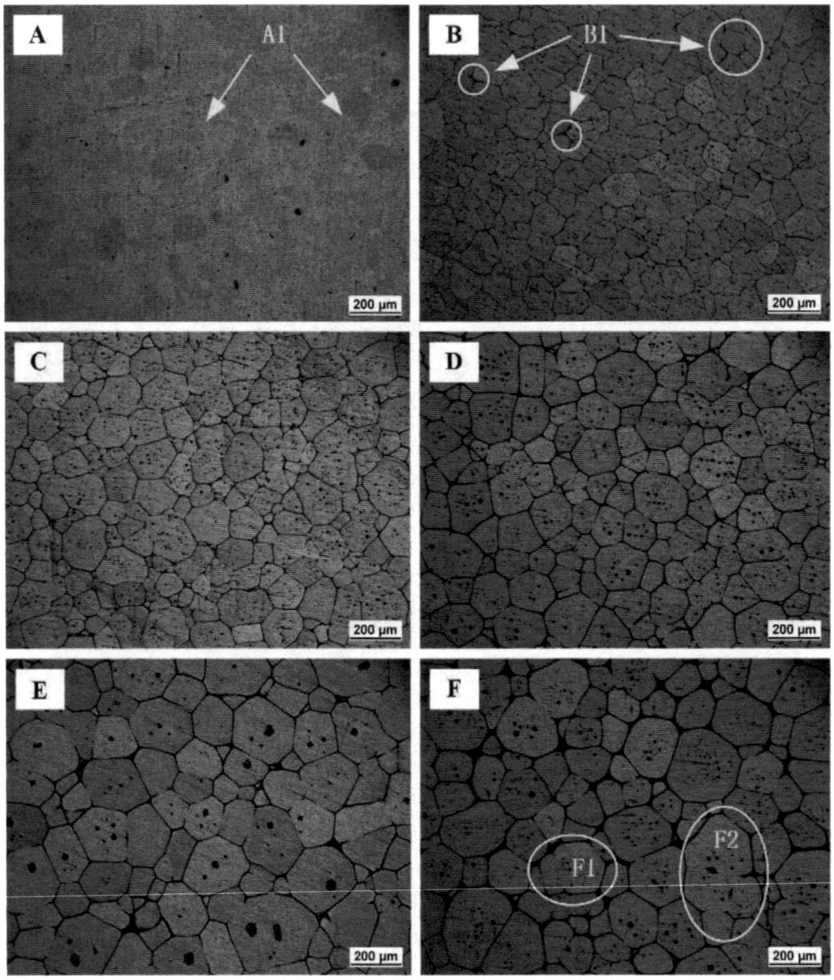

Figure 5. Microstructure of RFCUM-deformed alloy after SSIT at 630 °C for different times: (A) 1 min, (B) 3 min, (C) 5 min, (D) 10 min, (E) 15 min and, (F) 20 min.

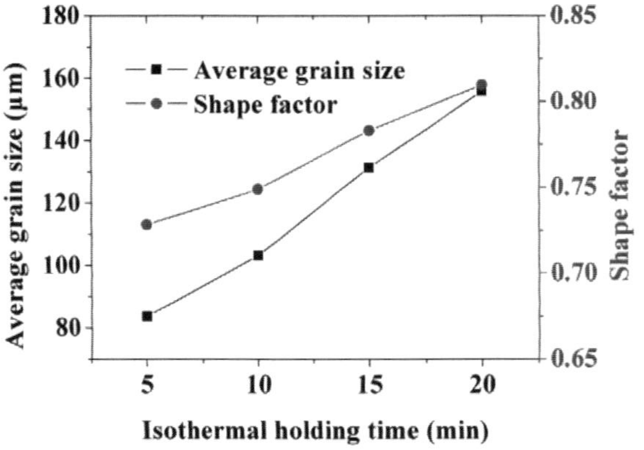

Figure 6. Variations of the average grain size and shape factor with the isothermal holding time

when the grain boundary energy is greater than twice the solid-liquid interface energy, which will lead to the formation of liquid film and fine solid grains (Figure. 5(C)). With further increasing the isothermal holding time, the grain coalescence occurs gradually, and then the average grain size is increased. At the same time, owing to the effect of interface curvature, the spheroidization degree of solid grain is improved (Figure. 5(D) ~ (E)). When the isothermal holding time is long enough to make the majority of solid grains separated with each other (Figure. 5(F)), the Ostwald ripening also starts to play a role on grain coarsening. From the results of Figure. 4~6, it can be concluded that, compared with the SSIT of as-received 6063 alloy, the RFCUC-RAP process in this study is an effective method to produce ideal semi-solid billets. The optimal isothermal holding temperature and time are 630 °C and 5~10 min, respectively; the average grain size is 83.80~103.15 μm and the shape factor is 0.73~0.75.

4 Conclusion

In this study, the microstructures of semi-solid 6063 Al alloy fabricated by semi-solid isothermal treatment (SSIT) and RFCUM-RAP process are investigated. The main conclusions are summarized as follows:

(1) Compared with the large and irregular solid grains produced by the SSIT process of as-received 6063 alloy, the RFCUC-RAP process is an effective method to obtain the ideal semi-solid billets with fine and spherical solid grains.

(2) For preparing the ideal semi-solid billet, the optimal isothermal holding temperature and time are 630 °C and 5~10 min, respectively. Furthermore, the average grain size is 83.80 ~103.15 μm and the shape factor is 0.73~0.75.

Acknowledgments

The authors are grateful to the National Natural Science Foundation of China (grant no. 51675414 and 51405378) and the Xi'an Science and Technology Plan Projects (grant no. CXY1514 (1)) for funding this study.

References

1. A. Bolouri, M. Shahmiri, C.G. Kang, J. Mater. Sci **47**, 3544 (2012)
2. Y. Xu, L. Hu, J. Jia, B. Xu, Mater. Charact **118**, 309 (2016)
3. Q. Chen, Z.D. Zhao, G. Chen, B. Wang, J. Alloy. Compd **632**, 190 (2015)
4. D.B. Spencer, R. Mehrabian, M.C. Flemings, Metall. Mater. Trans. B **3**, 1925 (1972)
5. Y.F. Wang, S.D. Zhao, C.Y. Zhang, R.G. Guan, Mater. Trans **58**, 176 (2017)
6. R. Haghayeghi, E.J. Zoqui, A. Halvaee, M. Emamy, J. Mater. Process. Tech **169**, 382 (2005)
7. P. Kaur, D.K. Dwivedi, P.M. Pathak, Int. J. Adv. Manuf. Tech **63**, 415 (2012)
8. N.V. Thuong, H. Zuhailawati, A.A. Seman, T.D. Huy, B.K. Dhindaw, J. Mater. Eng. Perform **24**, 2018 (2015)
9. Y. Murakami, K. Miwa, M. Kito, Mater. Trans **57**, 168 (2016)
10. B. Binesh, M. Aghaie-Khafri, Mater. Charact **106**, 390 (2015)
11. Y. Meng, S. Sugiyama, J. Yanagimoto, J. Mater. Process. Tech **214**, 87 (2014)
12. J.F. Jiang, Y. Wang, H.V. Atkinson, Mater. Charact **90**, 52 (2014)
13. H.V. Atkinson, K. Burke, G. Vaneetveld, Mat. Sci. Eng. A **490**, 266 (2008)
14. G.H. Yan, S.D. Zhao, S.Q. Ma, H.T. Shou, Mater. Charact **69**, 45 (2012)
15. J.F. Jiang, Z.M. Du, Y. Wang, S.J. Luo, Solid State Phenom **271**, 29 (2015)
16. B. Binesh, M. Aghaie-Khafri, Metals **6**, 42 (2016)

Investigation on the High-speed Precision Shearing of 5140 Steel Bars Pre-grooved by Laser

Yuanzhe Dong[1,], Wei Zhong[1], Yujian Ren[1], Shengdun Zhao[1,a] and Lijun Zhang[2]

[1] *School of Mechanical Engineering, Xi'an Jiaotong University Xi'an, Shanxi, China*

[2] *College of Electromechanical Engineering, China University of Petroleum, Qingdao, China*

Abstract. A new blanking method is proposed to improve the quality of metal bars, by introducing the laser pre-grooving and high-speed precision shearing technology. The working principle is introduced, and the calculation method of the laser groove depth is given. By means of finite element method, the elastic-plastic deforming, damage initiation and crack propagation behaviour of a 5140 steel bar under high-speed shearing are obtained; furthermore, the corresponded shear force-displacement property is analysed in detail. After that, high speed shearing experiments are done of laser grooved 5140 steel bars. The results show that the 5140 steel bar undergoes rapid elastic and little plastic deformation before crack, and micro cracks initially appears in the lateral area with the damage value of 0.53; then main vertical cracks form and extend along the shearing direction; the fracture section of the billet is flat and smooth, with a section inclination of 0.3°and a weight tolerance of 0.35%.

1 Introduction

Blanking of metal bars are the first process to produce most mechanical parts, such as gears and shafts in gearboxes, chain pins in motorcycles and bicycles, rings and rollers in bearings, precision pipe joints and standard metal parts (screws, bolts, nuts), and has enormous application in automobile, rail transit, and petrochemical equipment manufacturing industries.

Traditional blanking method use shearing dies with sharp blade to cut metal bar directly at a speed of 0.3~0.5m/s [1]. During the process, the area of the bar contacted to the blades first collapses, and then crack initiates and extents deviating from the vertical direction as the whole bar inclines due to bending. At last, billets are produced with horseshoe-shape cross sections and burrs, and need further section-machining process, together with problems such as badly wear of blades.

Several methods have been proposed to improve the blanking quality. High-speed precision shearing method enhances the brittleness of metals by increasing the loading speed [2], and research shows that the shearing velocity, when more than 5 m/s, can effectively improve the flatness and roundness of billets' section with a high efficiency. However, there are still collapse zones left at the contact area of bars, together with serious problems as wear and breakage of the blade. Low stress blanking method, based on the crack technology, firstly cut annular grooves on the surface of the bars to generate stress concentration, and then apply circumferential cyclic load to accelerate the initiation and extension of fatigue crack [3, 4], which can obtain billets with high quality section. Whereas, the blanking efficiency is relatively low(6~8 billets per minute); on the other hand, the stress grooves with

V shape profiles [5] are mainly processed by lathe or milling, which don't generate severe stress concentration, and grooving of hard materials is still difficult.

According to above reasons, this paper presents a novel combined blanking method of metal bars with laser pre-grooving and high-speed precision shearing, which firstly preprocesses ring-shape grooves on the surface of bars by laser, and then applies high-speed impact on the radially restrained metal bars to complete the separation. Take 5140 steel bars for example, which is widely used in manufacturing industry, shearing simulation and experiments are further done.

2 Working Principle of the Method

The whole working principle of the method is shown in Fig.1. Fig.1 (a) is the schematic diagram of the laser rotary grooving of metal bars. Suppose that a metal bar with diameter D rotates around the central axis Y at speed n, the laser beams form a series of circular contours with radius r on the bar surface. Compressed air are used as auxiliary gas to combust and blow slags, and the material is removed layer by layer to produce a certain annular groove with width w and depth h. And then equally spaced annular grooves are processed in sequence according to the length of the billets, through the linear interpolation movement of the laser head.

Fig.1 (b) shows the principle of the novel high-speed precision shearing. The pre-grooved metal bar is firstly radially constrained in the invented double-side shearing die [6] to avoid bending, and the first two circular grooves are placed between the floating shearing die and the two static shearing dies. Then the hammer impacts the floating block at a speed of 5~10m/s, so that the cracks propagates quickly and vertically from the tip of the first two grooves and two pieces of billets are obtained simultaneously. After that, the hammer is lifted by hydraulic fluid, and the floating block is reset by the spring. Then feeding goes on and the two billets are pushed out of the shearing die.

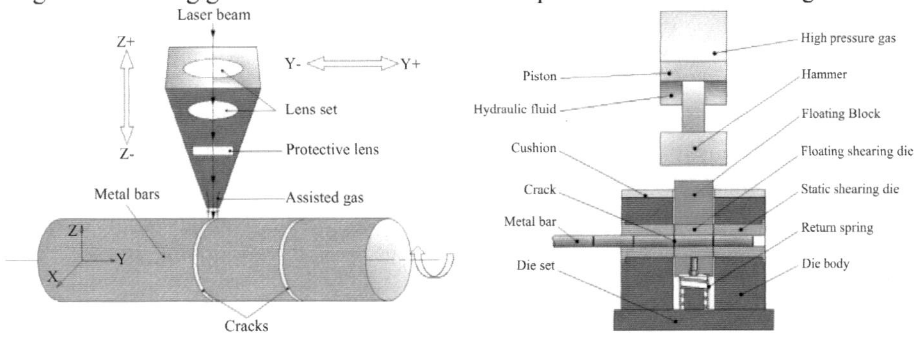

(a) laser rotary grooving (b) novel high-speed precision shearing

Figure.1. Schematic of the Combined Blanking Method

Fig.2 further illustrates the laser rotary grooving process. Fig.2 (a) shows the scanning trajectory of laser spots, and the overlap length δ of two adjacent holes determines the continuity of the annular groove. Suppose the number of spots in one circle is m, then m and δ can be calculated as follows:

$$m = (\frac{60}{n}) / (\frac{1}{f}) = \frac{60f}{n} \tag{1}$$

$$\delta = \begin{cases} 2r - \frac{\pi n D}{60f}, n < \frac{120rf}{\pi D} \\ 0, n \geq \frac{120rf}{\pi D} \end{cases} \tag{2}$$

Fig.2 (b) shows the pulse waveform of the Nd:Yag laser. The adjustable parameters include peak power P_k, pulse width τ, pulse frequency f. Its pulse intensity I_0 can be calculated by formula (3).

$$I_0 = P_k / \pi r^2 \tag{3}$$

 DOI: 10.1051/matecconf/201713601004

The width w of the groove is mainly determined by the radius r of the laser spots. The depth h is calculated as follow:

$$h = \int_0^t \frac{q(t)}{\rho[c(T_b - T_0) + L_f + L_v]} dt \tag{4}$$

c is the specific heat, T_0 and T_b is the initial and the boiling temperature of the material, L_f and Lv are the latent heat coefficients of fusion and vaporization [7], and q *(t)* is the time-dependent heat flux.

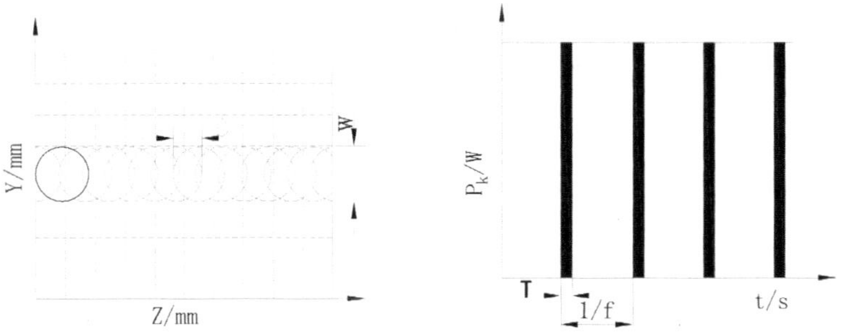

(a) Scanning trajectory (b) Pulse waveform

Figure 2. Scanning trajectory and pulse waveform of of the Nd:Yag laser

3 Simulation of the High–speed Shearing Process

In the finite element model, the mass of the hammer and the initial impact energy are set as 200kg and 8kJ separately, providing a loading speed of 8.9m/s. Take 5140 steel bar as example, the elasticity modulus E is 210GP, and poison's ratio v is 0.3. As an axial symmetry problem, half part of the grooved bar are analyzed. Its geometric parameters are as follows: billet length L=200mm, bar diameter D=26mm, depth of the groove h=1.51mm, width of the groove w=0.36mm. The actual radial clearance between the bar and the shearing dies is 0.5 mm.

The Normalized Cockcroft & Latham [8] formula is used as the fracture criterion, and the material cracks when the energy density exceeds the critical failure value. The expression is as follow.

$$D = \int_0^{\bar{\varepsilon}_f} \frac{\sup(\sigma_1, \sigma_2, \sigma_3)}{\sigma_{eq}} d\bar{\varepsilon}_p \geq D_c \tag{5}$$

In equation (5), $\sup(\sigma_1, \sigma_2, \sigma_3)$ is the maximum principal stress when it fractures, σ_{eq} is the equivalent stress, $\bar{\varepsilon}_p$ is the equivalent plastic strain, and $\bar{\varepsilon}_f$ is the critical equivalent plastic strain.

Fig.3 shows the total mechanics behavior of the 5140 steel bar at different shear strokes and the corresponded shear force. As shown in Fig.3 (a), at stroke of 0.6mm, no obvious collapse and bending occur in the contact area of the bar. The maximum damage value of 0.53 appears in the lateral area of the annular groove, which exceeds the critical fracture threshold of 0.52, so that the material in that zone begins to crack; at stroke of 4.2mm, the part of the bar in the floating shearing die bends a very small curvature of 0.69 °, and the maximum damage value is 0.785 at the circumferential lateral area of the annular groove, so that the crack extends rapidly; when the stroke is 7.8mm, the bar has completed separation. The fracture section of the billet is flat and smooth with no collapse, and the inclination of the fracture section is only 0.26 °, showing that the new blanking method can produce billets with high accuracy.

Fig.3 (b) shows the load-displacement curve during high-speed shearing. It has the characteristics of rapid elastic deformation, slight plastic deformation, instantaneous fracture and multiple-impact contacts. When the stroke is less than 0.8mm, the bar undergoes rapid elastic-plastic deformation to crack, and the maximum shear force is 232T; when the stroke is between 0.8mm and 3.8mm, the crack extends rapidly with static shear force of 34T; when the stroke exceeds 3.8mm, the bar fractures immediately, and the value of the shear force gradually decreases to zero. Then the floating block rebounds and contacts the hammer several times with the help of the return spring, resulting in sudden fluctuation of shear force.

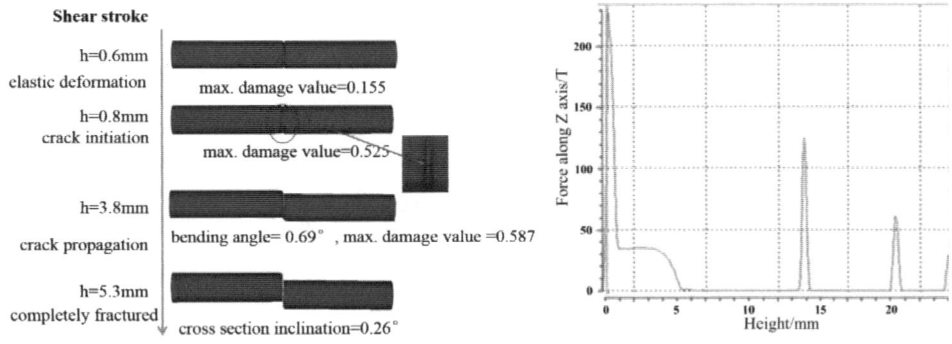

(a) fracture behavior (b) shear force-displacement curve

Figure 3. Fracture behavior of 5140 steel bar and corresponded shear force

4 Experimental results and discussion

Fig.4 (a) is the photo of the pre-grooved 5140 steel bars with diameter of 26mm. Fig.4 (b) shows one the the laser grooves on the bars, and the average width *w* and depth *h* of the grooves are 1.51mm and 0.24mm separately. The shearing experiment was done on the hydro-pneumatic hammer test bed as shown in Fig.4 (c). In this experiment, the pre-grooved 5140 steel bars were first placed in the invented shearing die, and then cropped at a speed of 8.9m/s.

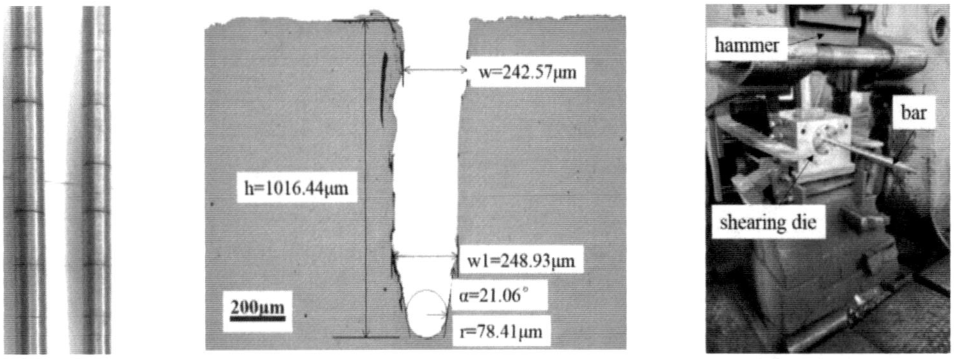

(a) pre-grooved 5140 steel bars (b) profile of the laser groove (c) laboratory setup

Figure 4. Experimetal materials and setup

4.1 Macro-fracture analysis of the 5140 billets

Fig.5 shows the profile and fracture appearance of a 5140 steel billet cropped with the new blanking method. The cross section of the billet is smooth and nearly round, and no horseshoe-shaped collapses produced; the section inclination is 0.3°, the weight tolerance is only 0.35%.

Fig.6 shows the result of the traditional warm blanking of the 5140 steel billet. The billet has an obvious horseshoe cross section, which is rough and out of round. The section inclination is up to 4.1°, and the weight tolerance is 2.5%, so that the extra plane blade cutting process is needed, leading the total time consumed to hundreds of seconds.

DOI: 10.1051/matecconf/201713601004

Compared with the traditional method, the new method can effectively improve the quality of the billets by more than 12 times and 6 times in accuracy of section inclination and weight tolerance separately.

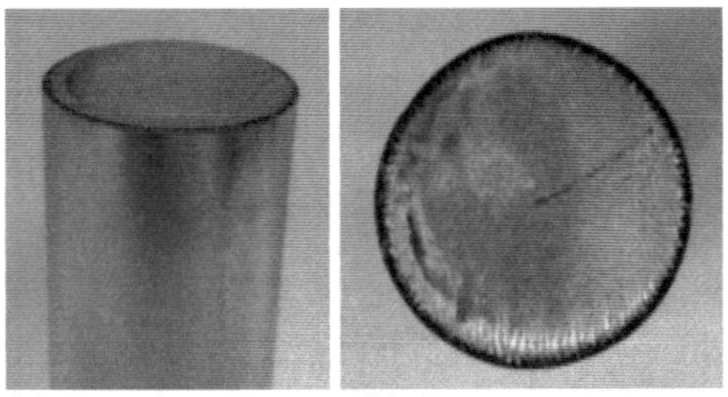

(a) front profile (b) fracture section

Figure 5. 5140 steel billet by the new method

(a) front profile (b) fracture section

Figure 6. 5140 steel billet by traditional blanking method

4.2 Micro-fracture analysis of the 5140 steel billets

The micro fracture morphology of the 5140 steel billet cropped by the new method is further analyses with the scanning electron microscopy (SEM) . As shown in Fig.7 (a), the width of the residual heat affected zone is less than 2mm. Fig.7 (b) magnifies the tip of the groove with 250 time, and it is obvious that the several micro cracks initiate at the tip. The opening angle of the groove increase to 76.1°, which reveals that the material at the bottom is first pulled to both sides, producing micro cracks under the tensile stress. After that, the micro cracks propagate and converge to form some larger vertical cracks and they extend along the shearing direction.

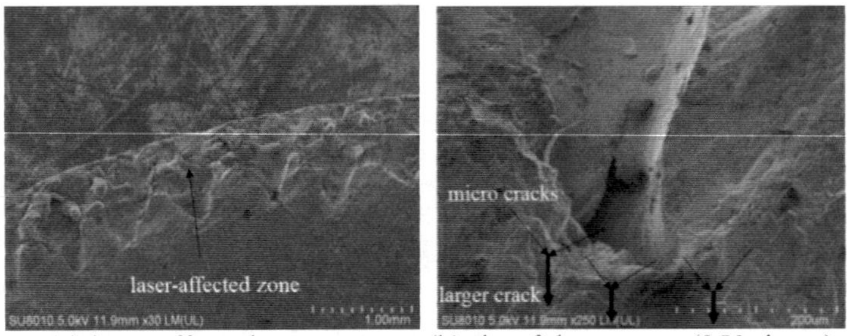

(a) laser-affected zone (b) tip of the groove (250 times)

Figure 7. Micro-fracture morphology of the 5140 steel billet

5 Conclusion

This paper presents a combined blanking method of metal bars with laser pre-grooving and high-speed precision shearing. Take 5140 steel bars as example, the finite element model is established and the results show that the shearing process has characteristic of rapid elastic deformation, slight plastic deformation and instantaneous fracture, and the fracture section of the billet is flat with no collapse. High-speed shearing experimental results further verify that method can efficiently improve the quality of 5140 steel billets with smooth and nearly round section. Specifically, compared to the traditional blanking method, the shearing surface inclination of 5140 steel billets is 0.3°, and the weight tolerance is only 0.35%, the precisions of which improve more than 12 times and 6 times separately. The blanked billets can be directly used for precision forming without extra turning process.

Acknowledgments

This work was supported by the National Natural Science Foundation of China (Grant No. 51335009, 51575532), Major National Science and Technology Project of China (Grant No. 2009ZX04005-031).

References

1. C. Gaudillière, N. Ranc, A. Larue, et al. *Investigations in high speed blanking: cutting forces and microscopic observations*. EDP Sciences,6,1221-1221 (2010)
2. J.H. Fu, Y.T. Li, B.F. Lei, et al. *Experiment research on the high-speed precision shearing of metal bar*. Journal of Mechanical Engineering, 41(5):154-156 (2005)
3. R.F. Zhao, S.D. Zhao, B. Zhong. *Experiment Study on Novel Precision Cropping Machinery Using Rotary Striking Action*. Applied Mechanics & Materials, 281,287-292 (2013)
4. B. Zhong, S.D. Zhao, R.F. Zhao, et al. *Numerical and experimetal investigation on the influence of main motor rotational frequency in fine-cropping*. ARCHIVE Proceedings of the Institution of Mechanical Engineers Part C Journal of Mechanical Engineering Science, 228(3),514-524 (2014)
5. B. Zhong, S.D. Zhao, R.F. Zhao, et al. *Investigation on the influences of clearance and notch-sensitivity on a new type of metal-bar non-chip fine-cropping system*. International Journal of Mechanical Sciences, 76(6), 144-151 (2013)
6. S.D. Zhao, Y.Z. Dong, F. Jing, et al. *A high-speed double - sided shearing method for radially restrained metal bars with low stress*. CN201610606879.3 (2016)
7. K. Kheloufik, E.H. Amara. *Numerical Investigation of the Effect of Some Parameters on Temperature Field and Kerf Width in Laser Cutting Process*. Physics Procedia, 39(7),872-880 (2012)
8. A. Reyes, M. Eriksson, O.G. Lademo, et al. *Assessment of yield and fracture criteria using shear and bending tests*. Materials & Design, 30 (3): 596-608 (2009)

Modification of the Lagrange Interpolating Polynomial Scheme for Using with the Finite Difference Method

Uthai Prasopchingchana[1,a]

[1]*Department of Mechanical Engineering, Burapha University, Chonburi 20131, Thailand*

Abstract. Modification of the Lagrange interpolating polynomial (LIP) scheme for using with the finite difference method is proposed. Merits of the modified LIP scheme used with the finite difference method for problem solving are facile to discretize equations, and fast to obtain solutions. Verification of the modified LIP scheme was performed by comparison of the solutions computed from the modified LIP scheme with the analytical solutions of a heat conduction problem. The verification gives credence to the modified LIP scheme for correctness of the solutions. In addition, the comparison results of solution accuracy and computational time of problem solving between using of the finite difference method with the modified LIP scheme and the finite volume method with the LIP scheme are exhibited.

1 Introduction

Most scientific and engineering problems are defined by partial difference equations. There are many numerical methods for solution determination of the problems such as the finite difference method (FDM), the finite volume method (FVM), the finite element method, the boundary element method and the lattice Boltzmann method etc. The FDM is not complicated, easy to understand and develop codes, and has low cost in memory and computational time. Problem solving by using high-order schemes is a means which high accurate solutions are desired. But, using the FDM with existing high-order schemes and non-uniform grids is quite burdensome.

In the last decade, there were many authors devising new high-order schemes not difficult to solve problems by using with the FDM and non-uniform grids. Filbet and Prouveur [1] developed a temporal high-order scheme for backward semi-Lagrangian methods and applied to transport equations for plasma physics problems. In addition, Chen et al. [2], Zhao and Deng [3], Wang et al. [4], Baeza et al. [5], and Jha and Kumar [6] constructed spatial high-order schemes for using with the FDM and non-uniform grids, and performed verification of their schemes with different problems.

Recently, the Lagrange interpolating polynomial (LIP) scheme for using with the FVM was proposed by Prasopchingchana and Manewattana [7]. The LIP scheme is easily modified for using with the FDM, called the modified LIP scheme.

An aim of this article is to propose a modified LIP scheme for using with the FDM. The modified LIP scheme is a high-order scheme used for discretization of the first and second derivatives of variables with respect to both space and time. Using the modified LIP scheme with the FDM to solve problems simulated by using non-uniform grids is facile, and solution convergence of the using is fast.

[a] Corresponding author : uthai@buu.ac.th

DOI: 10.1051/matecconf/201713601005

2 Modification of the LIP scheme for using with the finite difference method

From **Figure 1.**, the first derivative values of variables with respect to time can be approximated by using the modified LIP scheme and can be expressed as

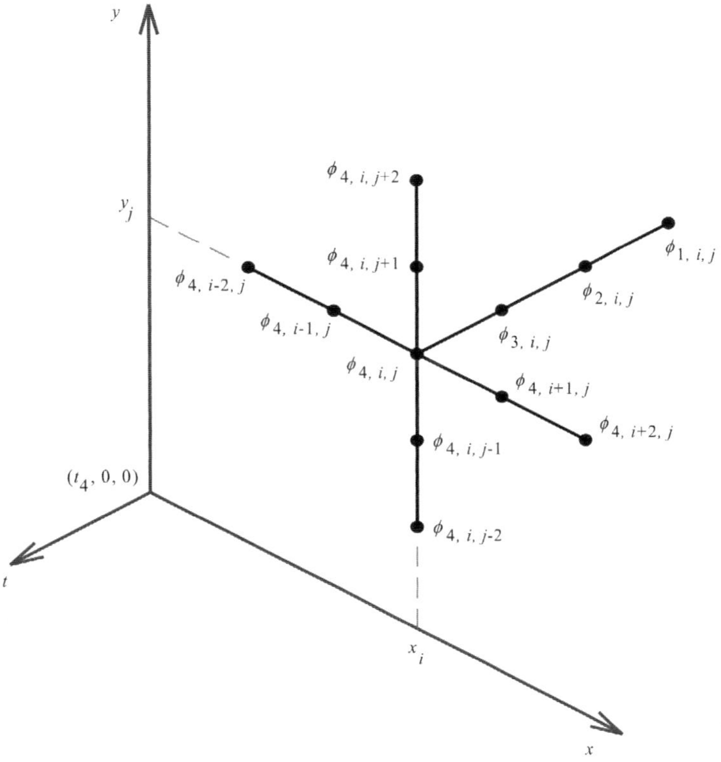

Figure 1. Schematic details of the modified LIP scheme for using with FDM.

$$\frac{d\phi_{4,i,j}}{dt} = \sum_{nt=1}^{4}\left(DLT_{nt}\,\phi_{nt,i,j}\right) \tag{1}$$

where ϕ, t and DLT are a variable, time and the first temporal derivative coefficients of the modified LIP scheme. The first temporal derivative coefficients in Equ. (1) can be computed from

$$DLT_{nt} = \frac{d}{dTT}\left(\prod_{\substack{m=1 \\ m\neq nt}}^{4}\frac{(TT - TT_m)}{(TT_{nt} - TT_m)}\right) \tag{2}$$

where $TT_{nt} = t_{nt} - t_4$, $TT_m = t_m - t_4$ and $TT = 0$

The first and second derivative values of variables with respect to space in the horizontal direction can be approximated by using the modified LIP scheme and can be written as

$$\frac{d\phi_{4,i,j}}{dx} = \sum_{nx=1}^{5}\left(DLX_{nx,i}\,\phi_{4,i-nxh+nx,j}\right) \tag{3a}$$

$$\frac{d^2\phi_{4,i,j}}{dx^2} = \sum_{nx=1}^{5}\left(DDLX_{nx,i}\,\phi_{4,i-nxh+nx,j}\right) \tag{3b}$$

where x, DLX, $DDLX$ and nxh are space in the horizontal direction, the first and second spatial derivative coefficients of the modified LIP scheme and a node constant value in the horizontal direction, respectively. The first and second spatial derivative coefficients in Equ. (3a) and (3b) can be calculated from

$$DLX_{nx,i} = \frac{d}{dX}\left(\prod_{\substack{ii=1\\ii\neq nx}}^{5}\frac{(X-X_{ii})}{(X_{nx}-X_{ii})}\right) \tag{4a}$$

$$DDLX_{nx,i} = \frac{d^2}{dX^2}\left(\prod_{\substack{ii=1\\ii\neq nx}}^{5}\frac{(X-X_{ii})}{(X_{nx}-X_{ii})}\right) \tag{4b}$$

where $X_{nx} = x_{i-nxh+nx} - x_i$, $X_{ii} = x_{i-nxh+ii} - x_i$ and $X = 0$. And the node constant value in the horizontal direction is $nxh = 1$ for $i = imin$, $nxh = 2$ for $i = imin + 1$, $nxh = 3$ for $imin + 2 \leq i \leq imax - 2$, $nxh = 4$ for $i = imax - 1$ and $nxh = 5$ for $i = imax$ where $imin$ and $imax$ are the minimum and maximum numbers of nodes in the horizontal direction.

Equ. (2), (4a) and (4b) can be simplified by changing to be algebraic forms. The algebraic forms of the first and second, temporal and spatial derivative coefficients of the modified LIP scheme are shown in **Table 1.**

Table 1. The first and second, temporal and spatial derivative coefficients of the modified LIP scheme.

First derivative coefficient	Second derivative coefficient	Remark
$DLT_{nt} = \dfrac{\sum_{\substack{m=1\\m\neq nt}}^{4}\left(\prod_{\substack{mm=1\\mm\neq m\\mm\neq nt}}^{4}(-TT_{mm})\right)}{\prod_{\substack{m=1\\m\neq nt}}^{4}(TT_{nt}-TT_m)}$	-	$TT_{nt} = t_{nt} - t_4$ $TT_m = t_m - t_4$ $TT_{mm} = t_{mm} - t_4$
$DLX_{nx,i} = \dfrac{\sum_{\substack{ii=1\\ii\neq nx}}^{5}\left(\prod_{\substack{iii=1\\iii\neq ii\\iii\neq nx}}^{5}(-X_{iii})\right)}{\prod_{\substack{ii=1\\ii\neq nx}}^{5}(X_{nx}-X_{ii})}$	$DDLX_{nx,i} = \dfrac{2\sum_{\substack{ii=1\\ii\neq nx}}^{5}\left(X_{ii}\left(\sum_{\substack{iii=1\\ii+iii\neq nx}}^{5-ii}(X_{ii+iii})\right)\right)}{\prod_{\substack{ii=1\\ii\neq nx}}^{5}(X_{nx}-X_{ii})}$	$X_{nx} = x_{i-nxh+nx} - x_i$ $X_{ii} = x_{i-nxh+ii} - x_i$ $X_{iii} = x_{i-nxh+iii} - x_i$
$DLY_{ny,j} = \dfrac{\sum_{\substack{jj=1\\jj\neq ny}}^{5}\left(\prod_{\substack{jjj=1\\jjj\neq jj\\jjj\neq ny}}^{5}(-Y_{jjj})\right)}{\prod_{\substack{jj=1\\jj\neq ny}}^{5}(Y_{ny}-Y_{jj})}$	$DDLY_{ny,j} = \dfrac{2\sum_{\substack{jj=1\\jj\neq ny}}^{5}\left(Y_{jj}\left(\sum_{\substack{jjj=1\\jj+jjj\neq ny}}^{5-jj}(Y_{jj+jjj})\right)\right)}{\prod_{\substack{jj=1\\jj\neq ny}}^{5}(Y_{ny}-Y_{jj})}$	$Y_{ny} = y_{j-nyh+ny} - y_j$ $Y_{jj} = y_{j-nyh+jj} - y_j$ $Y_{jjj} = y_{j-nyh+jjj} - y_j$

DOI: 10.1051/matecconf/201713601005

3 Verification of the modified LIP scheme

Verification of the modified LIP scheme for using with the FDM must be performed for confidence that the modified LIP scheme works correctly. Comparison of the solutions computed from the modified LIP scheme with the analytical solutions of a "heat conduction in rectangular plates with boundary temperatures specified" problem was adopted for the verification. The analytical solutions reported by Beck et al. [8] are the well-known solutions of the problem. An in-house code was developed based on how to solve the problem by using the modified LIP scheme with the FDM.

For the problem, interior temperatures in the rectangular plates with the specified boundaries as shown in **Figure 2.** can be calculated from

$$\rho c \frac{\partial T}{\partial t} - k\left(\frac{\partial^2 T}{\partial x^2} + \frac{\partial^2 T}{\partial y^2}\right) = 0 \tag{5}$$

where ρ, c, k and T are density, specific heat, thermal conductivity, and temperatures, respectively.

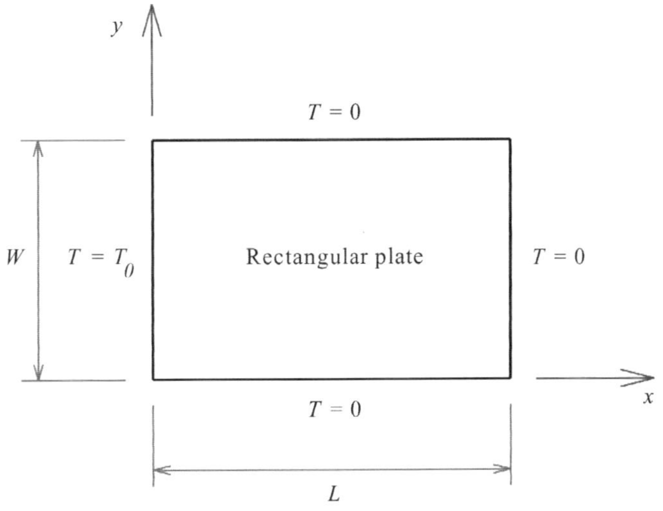

Figure 2. Schematic details of a "heat conduction in rectangular plates with boundary temperatures specified" problem for verification.

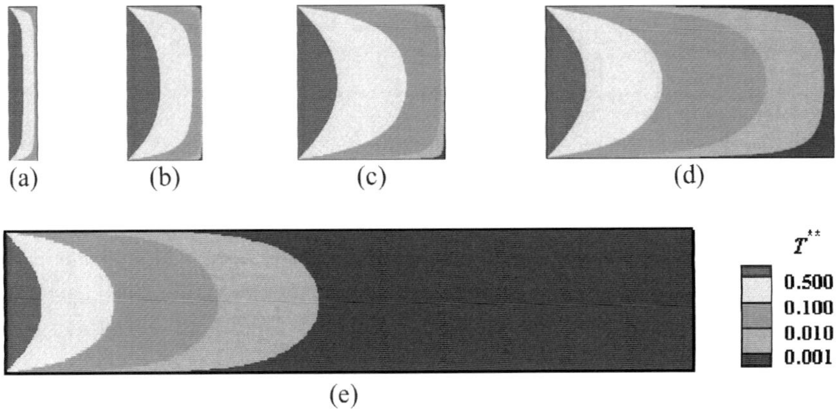

Figure 3. Dimensionless temperature contours of the solutions of a "heat conduction in rectangular plates with boundary temperatures specified" problem which are computed from the modified LIP scheme, (a) case 1, (b) case 2, (c) case 3, (d) case 4 and (e) case 5.

Table 2. Comparison of the solutions computed from the modified LIP scheme with the analytical solutions of a "heat conduction in rectangular plates with boundary temperatures specified" problem.

Case	Author	x^{**}	y^{**}	L^{**}	$T(x^{**},y^{**})/T_0$ (Accuracy, %)	$q''_x(x^{**},y^{**})L/kT_0$ (Accuracy, %)	$q''_y(x^{**},y^{**})L/kT_0$ (Accuracy, %)
1	[8]	0.10	0.25	0.20	0.4874535168	0.9992238948	-0.0393751511
	Present work	0.10	0.25	0.20	0.4875203277 (99.9862938931)	0.9992157526 (99.9991851476)	-0.0391183171 (99.3477256777)
2	[8]	0.25	0.25	0.50	0.3640566638	0.9169912516	-0.3798302130
	Present work	0.25	0.25	0.50	0.3641939947 (99.9622776030)	0.9166666901 (99.9646058237)	-0.3790895581 (99.8050036899)
3	[8]	0.50	0.25	1.00	0.1820283319	0.6387957290	-0.5371610386
	Present work	0.50	0.25	1.00	0.1822103823 (99.8999878766)	0.6387135936 (99.9871421495)	-0.5374608753 (99.9441812271)
4	[8]	1.00	0.25	2.00	0.0388578672	0.2453678480	-0.2435418264
	Present work	1.00	0.25	2.00	0.0389736704 (99.7019826142)	0.2453441945 (99.9903599839)	-0.2442754813 (99.6987560983)
5	[8]	2.50	0.25	5.00	0.0003495056	0.0054900240	-0.0054900207
	Present work	2.50	0.25	5.00	0.0003561433 (98.1008315747)	0.0054902856 (99.9952349935)	-0.0055946527 (98.0941419766)

Note

$$\text{Accuracy} = \left(1 - \left|\frac{[8] - \text{Present work}}{[8]}\right|\right) \times 100$$

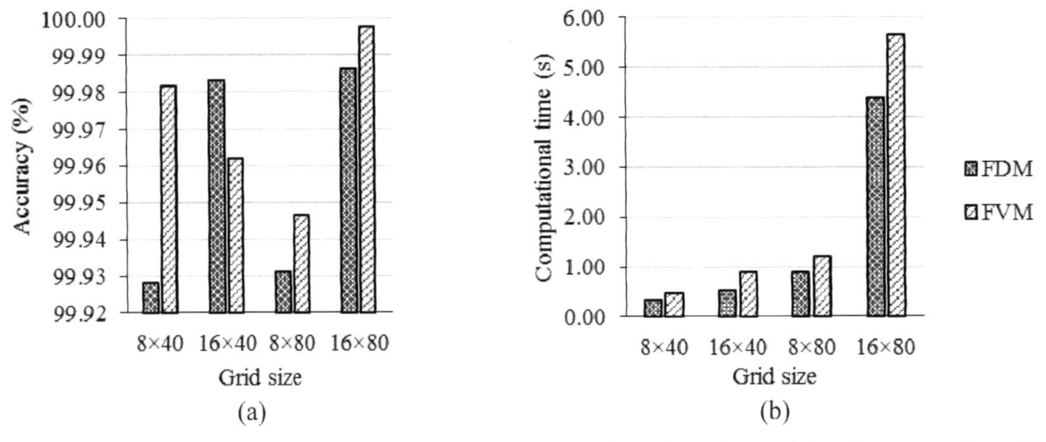

Figure 4. Comparison of solution accuracy and computational time of problem solving between using of the FDM with the modified LIP scheme and the FVM with the LIP scheme at the different grid sizes for case 1, (a) solution accuracy and (b) computational time.

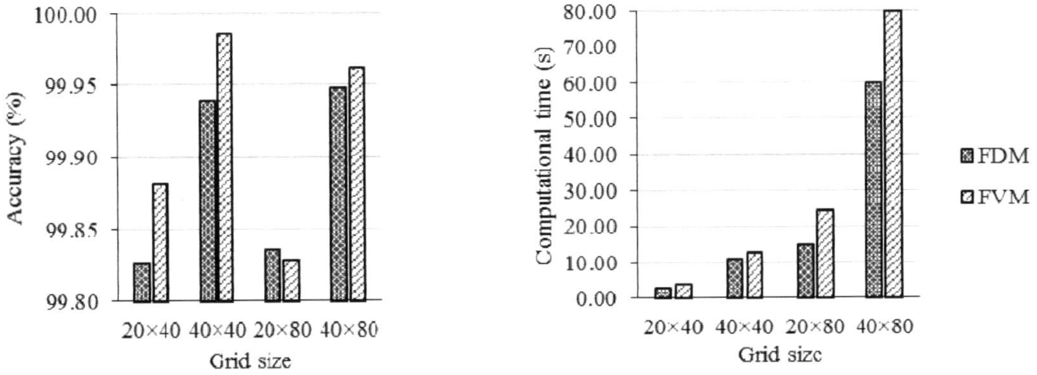

(a) (b)

Figure 5. Comparison of solution accuracy and computational time of problem solving between using of the FDM with the modified LIP scheme and the FVM with the LIP scheme at the different grid sizes for case 2, (a) solution accuracy and (b) computational time.

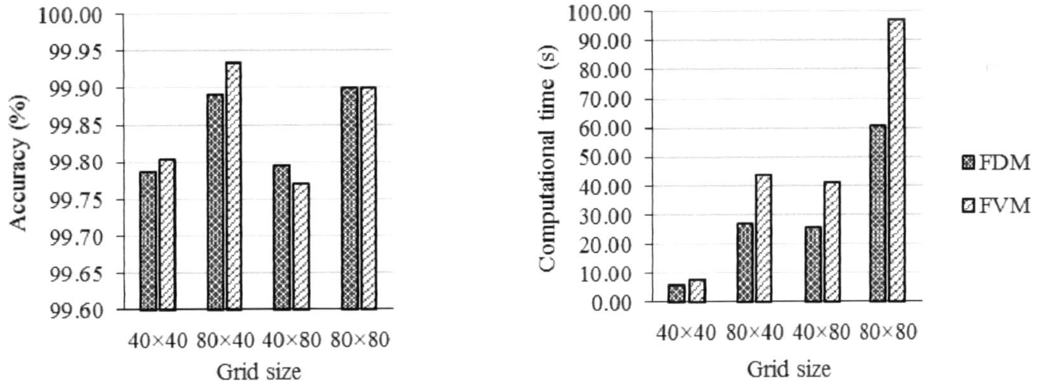

(a) (b)

Figure 6. Comparison of solution accuracy and computational time of problem solving between using of the FDM with the modified LIP scheme and the FVM with the LIP scheme at the different grid sizes for case 3, (a) solution accuracy and (b) computational time.

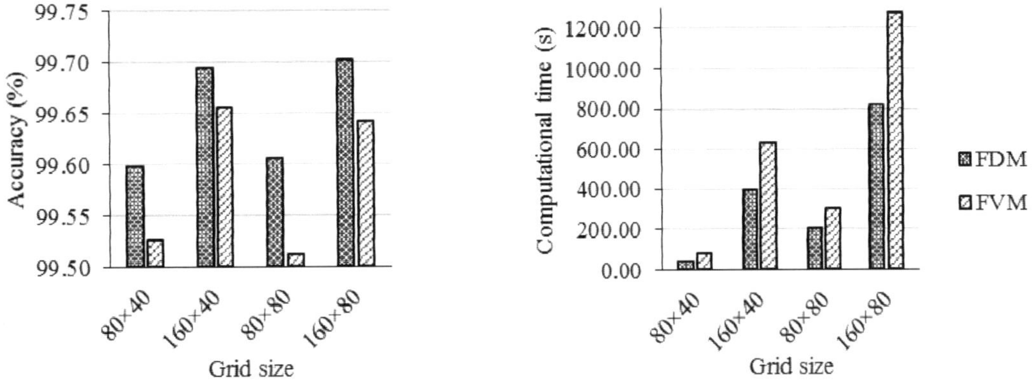

(a) (b)

Figure 7. Comparison of solution accuracy and computational time of problem solving between using of the FDM with the modified LIP scheme and the FVM with the LIP scheme at the different grid sizes for case 4, (a) solution accuracy and (b) computational time.

DOI: 10.1051/matecconf/201713601005

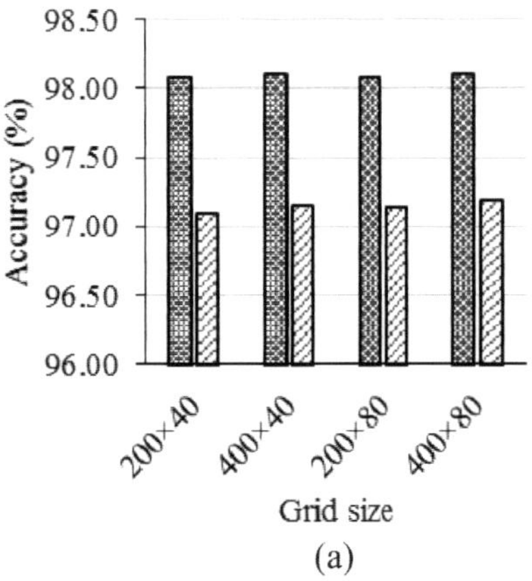

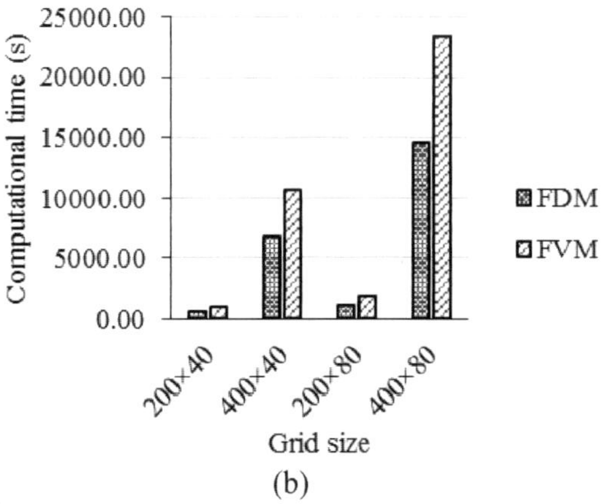

Figure 8. Comparison of solution accuracy and computational time of problem solving between using of the FDM with the modified LIP scheme and the FVM with the LIP scheme at the different grid sizes for case 5, (a) solution accuracy and (b) computational time.

The first temporal and second spatial derivative values of the temperatures in Equ. (5) can be discretized by using the modified LIP scheme.

$$\rho c\left(\sum_{nt=1}^{4}\left(DLT_{nt}\,T_{nt,i,j}\right)\right)-k\left(\left(\sum_{nx=1}^{5}\left(DDLX_{nx,i}\,T_{4,i-nxh+nx,j}\right)\right)+\left(\sum_{ny=1}^{5}\left(DDLY_{ny,j}\,T_{4,i,j-nyh+ny}\right)\right)\right)=0 \tag{6}$$

Equ. (6) can be rearranged to be

$$T_{4,i,j}=\frac{\left(\begin{array}{l}-\rho c\left(\sum_{nt=1}^{3}\left(DLT_{nt}\,T_{nt,i,j}\right)\right)\\ +k\left(\sum_{nx=1}^{5}\left(DDLX_{nx,i}\,T_{4,i-nxh+nx,j}\right)\right)\Bigg|\begin{array}{l}\text{when}\\ nx\neq nxh\end{array}\\ +k\left(\sum_{ny=1}^{5}\left(DDLY_{ny,j}\,T_{4,i,j-nyh+ny}\right)\right)\Bigg|\begin{array}{l}\text{when}\\ ny\neq nyh\end{array}\end{array}\right)}{\left(\rho c\,DLT_4-k\,DDLX_{nxh,i}-k\,DDLY_{nyh,j}\right)} \tag{7}$$

The Gauss-Seidel iterative method with the non-uniform grids were used to solve the equation system of Equ. (7). Fine grids were drawn near the boundaries of the rectangular plates, due to the steep temperature gradients. And coarse grids were applied in an interior area of the rectangular plates. A spatial ratio of the coarse grids to the fine grids was 1.05. A relative error of the code was less than or equal to 10^{-10}, and a time step value of the code was 0.01 s.

Table 2. shows comparison of the solutions computed from the modified LIP scheme with the analytical solutions of the "heat conduction in rectangular plates with boundary temperatures specified" problem. There are 5 cases of the difference of the dimensionless parameters for the comparison. For the present work, the modified LIP scheme which is the spatial 4th order (a 9 nodes, 2D stencil) and the 4 time steps is adopted for discretization. The dimensionless parameters displayed in **Table 2.** can be calculated from $x^{**}=x/W$, $y^{**}=y/W$ and $L^{**}=L/W$. The accuracies of the solutions computed from the modified LIP scheme can add credence to the modified LIP scheme.

Figure 3. illustrates dimensionless temperature contours of the solutions of the "heat conduction in rectangular plates with boundary temperatures specified" problem which are computed from the modified LIP scheme for 5 cases. The contour values in **Figure 3.** are specified according to the contour values shown in [8], and the contour pattern displayed in **Figure 3.** is similar to the contour pattern exhibited in [8].

Figure 4.-8. display comparison of the solution accuracies and computational time of the problem solving between using the FDM with the modified LIP scheme and using the FVM with the LIP scheme at different grid sizes for 5 cases. CPU time of the computer is adopted to be the computational time of the code. Specification of the computer used for code computation is "Intel ® Core ™ i5-4460 CPU @ 3.20 GHz 3.20 GHz". For the comparison of the solution accuracy, it can not specify which of the FDM or the FVM can give more accurate solutions. But, for the comparison of the computational time, the computational time of the problem solved by using the FDM with the modified LIP scheme is less than the computational time of the problem solved by using the FVM with the LIP scheme in all of the grid sizes and cases.

4 Conclusion

Modification of the LIP scheme for using with the FDM is presented. And, verification of the modified LIP scheme was performed. Advantages of the modified LIP scheme for using with the FDM are that it is a high-order scheme, easy to understand and develop codes, and spends less computational time for problem solving than by using the FVM with the LIP scheme.

Acknowledgment

The author wishes to acknowledge the Faculty of Engineering, Burapha University for the financial support to attend D2ME 2017.

References

1. F. Filbet, C. Prouveur, J. Comput. Appl. Math., **303**, 171 (2016)
2. W. Chen, J. C. Chen, E. Y. Lo, Comput. Fluids, **101**, 273 (2014)
3. L. Zhao, W. Deng, Adv. Comput. Math., **42**, 425 (2016)
4. S. Wang, K. Virta, G. Kreiss J. Sci. Comput., **68**, 1002 (2016)
5. A. Baeza, P. Mulet, D. Zorío, J. Sci. Comput., **69**, 170 (2016)
6. N. Jha, N. Kumar, Adv. Differ. Equ. **64**, 1 (2017)
7. U. Prasopchingchana, T. Manewattana, Eng. J., **19**, 133 (2015)
8. J. V. Beck, N. T. Wright, A. Haji-Sheikh, K. D. Cole, D. E. Amos, Int. J. Heat Mass Transf., **51**, 4676 (2008)

MATEC Web of Conferences **136**, 01006 (2017) DOI: 10.1051/matecconf/201713601006
D2ME 2017

The Evaluation of Surface Integrity During Machining of Inconel 718 with Various Laser Assistance Strategies

Szymon Wojciechowski[1,a], Damian Przestacki[1] and Tadeusz Chwalczuk[1]

[1]*Poznan University of Technology, Institute of Mechanical Technology, Piotrowo 3, Poznan 60-965, Poland*

Abstract. The paper is focused on the evaluation of surface integrity formed during turning of Inconel 718 with the application of various laser assistance strategies. The primary objective of the work was to determine the relations between the applied machining strategy and the obtained surface integrity, in order to select the effective cutting conditions allowing the obtainment of high surface quality. The carried out experiment included the machining of Inconel 718 in the conventional turning conditions, as well as during the continuous laser assisted machining and sequential laser assistance. The surface integrity was evaluated by the measurements of machined surface topographies, microstructures and the microhardness. Results revealed that surface integrity of Inconel 718 is strongly affected by the selected machining strategy. The significant improvement of the surface roughness formed during machining of Inconel 718, can be reached by the application of simultaneous laser heating and cutting (LAM).

1 Introduction

Super alloys based on the nickel, including Inconel 718 are materials which are characterized by the unique properties, as: excellent strength at the elevated temperatures, increased thermal conductivity, as well as resistance to oxidation and corrosion at the high temperatures. Thus, these materials are being widely applied in the aircraft, turbine and nuclear power plant applications. According to Ezugwu [1] the aerospace engine component is made of nickel-based alloys in about 50 % of its weight. Nevertheless, these excellent properties of super alloys significantly affect their machinability. The primary problems occurring during machining of Inconel alloys include the: high values of forces, the concentration of heat, tool fracture induced by plastic deformation during cutting with higher speeds, the decreased tool life and the work hardening phenomenon [2]. Therefore the nickel based super alloys are the materials which are difficult-to-cut.

Literature survey shows that many researches are focused on improvement of machinability by the application of high-pressure cooling [3], plasma enhanced machining [4, 5], laser assisted machining (LAM) [6-8]. However, according to Venkatesan *et al.* [9], the laser assisted machining is an approach which gives increased benefits without compromising quality, in comparison to other novel technologies. The LAM technology is based on the application of high power laser, which heats the workpiece before the initiation of chip removal induced by a tool [10]. The aim of the process is to reduce the cutting forces by an increase of machining temperature, which leads to the reduction of the material's strength. The main advantages of the LAM are: improvement in surface roughness, decrease in specific cutting energy, growth of material removal rates, improvement of process stability and reduction of tool's catastrophic fracture and wear [11-13]. The application of LAM during machining of Inconel significantly enhances its machinability. Nevertheless, the improvement of machined surface integrity requires the selection of the appropriate cutting conditions and strategies. The machined surface integrity is affected by the kinematic-geometric parameters, process stability [14] and vibrations [15,16], plastic-elastic deformations of work material induced by the ploughing mechanism [17], as well as the thermal phenomena in the cutting zone [18]. However, in case of difficult-to-cut materials, the effective application of LAM can result in improvement of machined surface quality. According to Przestacki *et al.* [19], the application of LAM during turning of MMC composite induces the sedimentation of hard SiC particles and thus causes the decline in surface roughness. In addition, the LAM of Inconel 718 in the appropriately selected temperature of the cutting zone can contribute to the decrease of tool wear rate and thus to the improvement of surface roughness [20].

The literature survey reveals that LAM of difficult-to-cut materials has been extensively investigated by many researchers [21]. However, the problems of machined surface integrity formed during LAM of Inconel 718, considering the cutting strategy have not been investigated sufficiently. Therefore this paper is focused on the relations between the applied laser assisted machining strategy and the obtained surface integrity. The conducted machining tests involved the following strategies: conventional turning (without laser assistance), sequential process with laser heating prior to turning and simultaneous laser heating and cutting (LAM). The obtained results can be applied for the selection of the effective cutting conditions allowing the improvements in surface finish.

[a] Corresponding author : sjwojciechowski@o2.pl

2 Experimental details

2.1 Workpiece and cutting tool

The Inconel 718 super alloy was applied as the workpiece. Its chemical composition is depicted in Table 1. The workpieces were provided as a cylinders with a diameter of 44 mm. The sample's microstructure consists mainly of γ solid solution grains, as well as the visible intermetallic phases, probably γ' and γ''.

Table 1. Chemical composition of Inconel 718 [%]

Element	Content
Ni	50-55
Cr	17-21
Nb	4.75-5.5
Mo	2.8-3.3
Ti	0.65-1.15
Al	0.2-0.8
Co	1
Mn	0.35
Si	0.35
Cu	0.3
C	0.08
P	0.015
S	0.015
B	0.006
Fe	Bal.

The samples dedicated to LAM were covered with a gouache layer in order to increase the absorption of laser's beam. The conventional and laser assisted turning tests (LAM) were carried out with the application of the coated sintered carbide KC5510 inserts with the SNMG120408 MS symbol and the following geometry: tool major cutting edge angle κr = 45°, flank angle in the feed motion direction αf = 5°, rake angle in the feed motion direction – γf = -5° and tool radius rε = 0.8 mm.

2.2 Machining tests

The carried out experiments (see – Tab. 2) involved the conventional longitudinal turning process (without laser assistance, designated as S4), laser assisted machining (LAM) including the simultaneous laser heating and turning of the heated surface (designated as S1 and S3), and sequential process which consisted of the heating of workpiece, its cooling down and then, turning (designated as S2).

Table 2. Design of experiment

No	Laser power P [W]	Laser spot diameter [mm]	Depth of cut a_p [mm]	Feed f [mm/rev]	Cutting speed v_c [m/min]	Laser scanning speed v_l [mm/min]
S1	1200	1	0.25	0.2	5.8	5.8
S2					28	5.8
S3					28	28
S4	0				28	0

The laser heating experiments were made on LAM stand equipped with Trumpf CO2 molecular laser integrated with TUM 35D1 lathe. The laser had the maximum power equaled to 2600 W and the radiation power density distribution type TEM01.

2.3 Inspection of surface integrity

The metallographic microsections of samples after machining were prepared in order to evaluate the microstructure and microhardness. These samples were mounted in the conductive resin, polished with the suspension, containing Al2O3, and ultimately etched in order to reveal their microstructure. The metallographic microsections of the machined samples were investigated with the application of a OPTA-TECH LAB-40 optical microscope and TESCAN scanning microscope, type VEGA TS 5135, equipped with an AWALON 4000 system. The hardness was determined with the application of Micromet II Buehler hardness meter.

The evaluation of surface topography was made with the application of T8000 stationary stylus surface profiler. The following settings were applied during the measurements: the sampling length equal to 0.8 mm, the measurement length equal to 4.0 mm. Based on the obtained surface profiles, the surface roughness parameters (Sa, Sz) were calculated with the use of dedicated software.

 DOI: 10.1051/matecconf/201713601006

3 Results and discussion

Figure 1 depicts the comparison of the Inconel 718 alloy's microstructure after various machining strategies. The analysis of SEM images shows that the application of distinct machining strategies has no significant effect on the obtained microstructure of samples. In all cases, the slightly lower grain sizes in the area near the surface layer (Fig. 1a) were found, in comparison to those contained in the base (Fig. 1b). It should be also noted that during machining with laser assistance (S1, S2, S3), the whole re-melted area has been probably totally removed by the cutting tool. Thus the SEM images presented in Fig. 1 are similar, independently on the applied machining strategy.

Despite the similar microstructure of samples, obtained after machining in various strategies, the distribution of microhardness along the radial direction reveals some differences (Fig. 2). In case of non-machined sample (M00), the microhardness oscillates around the 400 HV. However for the all machined samples, the decreasing trend of hardness values together with the growth of distance from the surface layer is observed. It is caused by the appearance of thermal phenomena, occurring in the shearing zone, resulting from the chip removal mechanisms. The variations in microhardness values, along the depth, reach up to 62%. It should be noted, that in case of conventional turning (S4) and simultaneous laser assisted machining (S1 and S3), the distribution of microhardness is similar. The lack of differences between these distinct machining strategies is probably caused by a relatively low depth of area re-melted by the laser beam. In example, during LAM of metal matrix composite with laser power P=1000 W, the depth of re-melted area is usually lower than 100 μm [19]. Furthermore, the increase of cutting and laser scanning speeds during LAM induces the slight decline in microhardness values in the range of radial depths to approx. 250 μm (see the hardness curves for the S1 and S3 in Fig. 2).

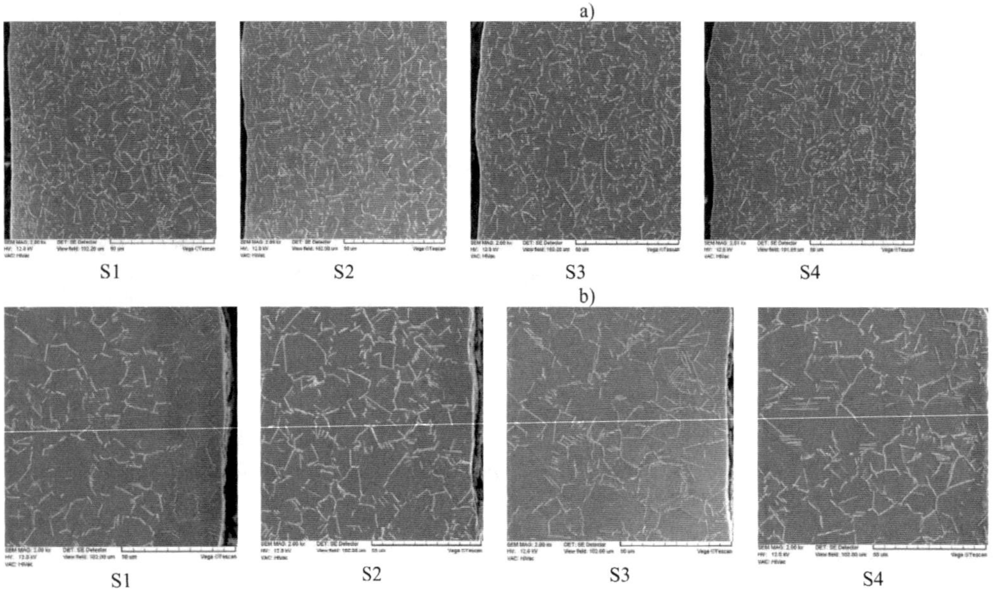

Figure 1. The SEM images of the Inconel 718 samples after various machining strategies: a) microstructure in the area of surface layer; b) microstructure in the area of base.

Figure 3 shows also that the lowest microhardness values, in the whole investigated range of sample's depth are found after the sequential machining process (S2). In this case the cooling time of workpiece is being extended in comparison to that appearing during LAM process, because of lack of heated chip removal action. Consequently, the higher amount of heat is transferred into the workpiece, which leads to the decrease of its microhardness. The surface topographies, obtained during turning in the range of investigated strategies are presented in figure 3. It can be observed that for the S2-S4 samples, the regular and constant distance between the formed irregularities is found. Its value corresponds to feed per revolution (f=0.2 mm/rev), and results from the kinematic-geometric model (so-called feed marks). Nevertheless, in case of S1 sample, machined in the LAM conditions with the lowest cutting speed and laser scanning speed, the distances between irregularities are variable, and thus the surface topography has a less regular form. This observation is strictly correlated with the lateral plastic flow of the overheated workpiece material, which in turn is attributed to the high heat transfer between the laser spot and workpiece's heated surface, occurring at the low cutting speed and laser scanning speed. Consequently the high heat transfer induces the growth of machined surface temperature and thus the intense softening of workpiece.

DOI: 10.1051/matecconf/201713601006

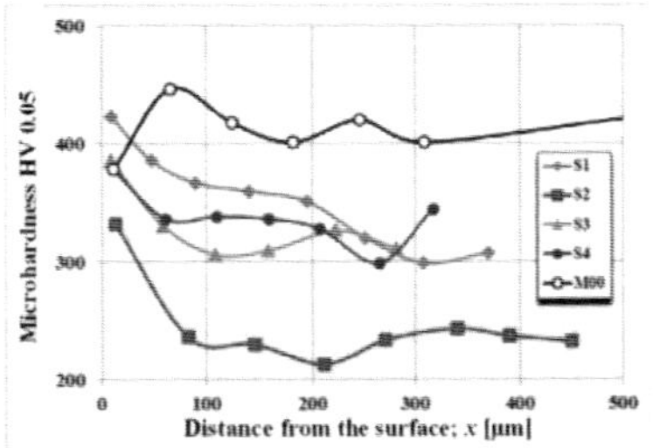

Figure 2. The distribution of microhardness in the investigated samples, along the radial direction from the surface layer.

In case of S4 sample, machined during conventional turning, some burrs are found on the peaks of formed irregularities. Their appearance can be related to the local brittle cracking of workpiece during material removal by the tool tip.

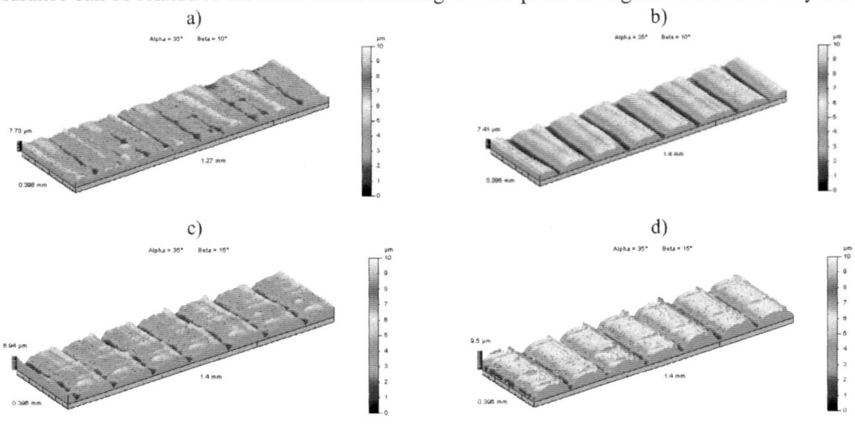

Figure 3. Surface topographies of Inconel 718 samples obtained during investigated turning strategies: a) S1; b) S2; c) S3; d) S4.

The analysis of surface roughness parameters (Figs. 4a,b) reveals also that the highest values of Sa and Sz are found for the S4 sample machined during conventional turning. The lowest surface roughness can be obtained during the laser assisted machining (LAM), including the simultaneous laser heating and turning of the heated surface (S1, S3). However, the application of sequential process enables the obtainment of surface roughness parameters ranked between the LAM and conventional machining.

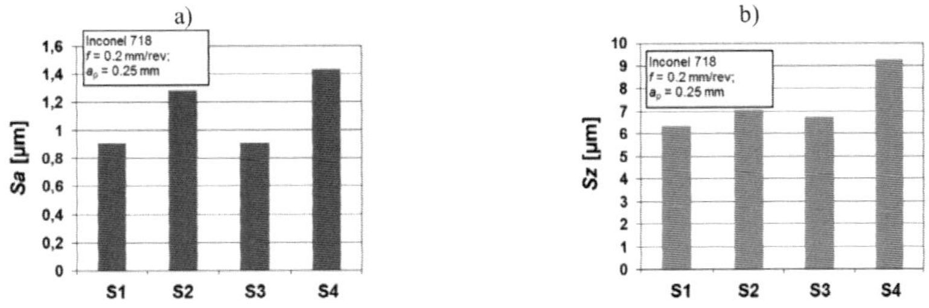

Figure 4. The comparison of surface roughness parameters obtained during turning with various strategies : a) Sa ; b) Sz.

4 Conclusions

The conducted research allowed to formulate the following conclusions:

- the distribution of microhardness in the Inconel 718 samples after turning is depended on the selected machining strategy. The lowest microhardness values in the whole investigated depth of sample can be reached during sequential process, which involves heating of workpiece, cooling down and then, turning;
- during LAM with the low cutting speed and laser scanning speed values, the overheating of the surface layer can be found, which results in the lateral plastic flow of the workpiece and thus formation of surface profile without the regular feed marks;
- in case of conventional turning of Inconel 718, the undesirable local brittle cracking of workpiece can occur, which affects the surface profile formation and thus the growth of surface roughness;
- the application of LAM improves the machined surface quality by the reduction of surface roughness parameters' values. The differences between the Sa values obtained during conventional turning and LAM reach up to 57%.

Acknowledgment

The presented research results, executed under the domestic project PBS of No 244445, were funded with grants for education allocated by the National Centre for Research and Development.

References

1. E.O. Ezugwu, Int. J. Adv. Manuf. Technol. **45** (2015)
2. D. Dudzinski, A. Devillez, A. Moufki, D. Larrouquere, V. Zerrouki, J. Vigneau, Int. J. Adv. Manuf. Technol. **44** (2004)
3. E. O. Ezugwu, J.Bonny, Tribol. Trans. **48** (2005)
4. C.E. Leshok, J. Kim, Y.C. Shin, Int. J. Mach. Tools Manuf. **4** (2001)
5. Z.Y. Wang, K.P. Rajurkar, J. Fan, S. Lei, Y.C. Shin, G. Petrescu, Int. J. Mach. Tools Manuf. **43** (2003)
6. P.A. Rebro, Y.C. Shin, F.P. Incropera, Int. J. Adv. Manuf. Technol. **44** (2004)
7. S. Lei, Y.C. Shin, F.P. Incropera, J. Manuf. Sci. Eng. Trans. ASME **123** (4) (2001)
8. M. Anderson, R. Patwa, Y.C. Shin, Int. J. Adv. Manuf. Technol. **48** (2006)
9. K. Venkatesan, R. Ramanujam, P. Kuppan, Optics & Laser Technology **78** (2016)
10. P. Dumitrescu, P. Koshy, J. Stenekes, MA. Elbestawi, Int. J. Mach. Tools Manuf. **46** (2006)
11. D. Przestacki, M. Jankowiak, Journal of Physics: Conference Series **483** (2014)
12. D. Przestacki, Procedia CIRP **14** (2014)
13. S. Sun, M. Brandt, M.S. Dargusch, Int. J. Mach. Tools Manuf. **50** (2010)
14. R. Rusinek, M. Wiercigroch, P. Wahi, Procedia CIRP **31** (2015)
15. S. Wojciechowski, P. Twardowski, M. Pelic, R.W. Maruda, S. Barrans, G. Krolczyk, Precision Engineering **46** (2016)
16. S. Wojciechowski, P. Twardowski, M. Pelic, Procedia CIRP **14** (2014)
17. S. Wojciechowski, R.W. Maruda, P. Nieslony, G.M. Krolczyk, International Journal of Mechanical Sciences **119** (2016)
18. D. Przestacki, T. Chwalczuk, S. Wojciechowski, Int J Adv Manuf Technol, DOI 10.1007/s00170-017-0035-5 (2017)
19. D. Przestacki, P. Szymanski, S. Wojciechowski, Composites Part A: Applied Science and Manufacturing **91** 1 (2016)
20. S. Azhdari Tadavani, R. Shoja Razavi, R. Vafaei, Optics & Laser Technology **87** (2017)
21. D. Przestacki, L. Marciniak-Podsadna, R. Majchrowski, Applied Surface Science **388** (2016)

 DOI: 10.1051/matecconf/201713601007

Ni-Zn Ferrite-graphene Nanohybrids: Synthesis and Characterization of Magnetic and Microwave Absorbing Properties

Yau Thim Ng[1], Wei Kong[2], Kim Yeow Tshai[1] and Ing Kong[1, a]

[1]*Department of Mechanical, Materials and Manufacturing Engineering, The University of Nottingham Malaysia, Jalan Broga 43500 Semenyih, Selangor, Malaysia*
[2]*Faculty of Applied Science and Foundation Studies, Infrastructure University Kuala Lumpur Unipark Suria, Jalan Ikram-Uniten, 43000 Kajang, Selangor, Malaysia.*

Abstract. An in-situ deposition technique was used in the synthesis of Ni-Zn ferrite-graphene (NZFG) nanohybrids. The XRD patterns revealed the presence of cubic spinel structure of Ni-Zn ferrite (NZF) nanoparticles with good crystallinity and small crystallite sizes. The SEM images showed NZF nanoparticles were uniformly deposited on graphene sheets. The effect of different loading amounts of NZF nanoparticles in the nanohybrids was also investigated by tuning the mass ratio of $FeCl_3$ and expanded graphite (EG). The magnetic measurements showed ferromagnetic behaviour with low coercivity. Improvements in saturation magnetization of the nanohybrids can be seen with increasing mass ratio of $FeCl_3$:EG. The microwave absorption properties were determined based on the measured relative complex permittivity and permeability. For the nanohybrids, the minimum reflection loss (R_L) obtained is -37.57 dB at 7.54 GHz and the absorbing bandwidth in which the R_L is less than -10 dB is 7.30 GHz when the NZF content was 79 wt·% at 7 mm thickness. The enhancement in the minimum R_L was due to the synergistic effect between NZF nanoparticles and graphene.

1 Introduction

In recent years, multifunctional nanocomposites comprised of graphene and magnetic nanoparticles have gained a significant amount of interest due to their combinations of properties that are potentially useful in a variety of applications [1,2]. Graphene, a wonder material having extraordinary electrical, thermal and mechanical properties [3], possesses large surface areas which can be used as a template for the attachment of magnetic nanoparticles.

Ni-Zn ferrite (NZF) is considered an excellent candidate of magnetic materials due to their high saturation magnetization, high electrical resistivity, good mechanical strength and chemical stability [4]. In terms of their synthesis methods, chemical methods are more advantageous over traditional milling methods in the preparation of ferrites due to its precise compositions, higher degree of homogeneity and better control of nanoparticle size. In particular, chemical co-precipitation is considered the most convenient method due to its simplicity and better control over the crystallite size and other properties of the materials. However, in many cases subsequent heating and annealing of the material is required to form the final product [4–7]. In light of works by [8–10], it has become possible to synthesize ferrites at room temperature.

In the present work, Ni-Zn ferrite-graphene (NZFG) nanohybrids with a Ni-Zn ferrite (NZF) composition of $Ni_{0.5}Zn_{0.5}Fe_2O_4$ were successfully synthesized through a facile in-situ deposition method at room temperature. The graphene sheets were produced by ultrasonication of expanded graphite (EG). The mass ratio of $FeCl_3$ to EG was varied from 5:1, 10:1 to 20:1 giving three different samples denoted as NZF5G1, NZF10G1 and NZF20G1. The microstructure, morphology and magnetic properties of the resulting nanohybrids were reported.

2 Experimental

2.1 Preparation of Ni-Zn ferrite graphene (NZFG) nanohybrids

NZFG nanohybrids were prepared by an in-situ deposition method involving the co-precipitation of NZF nanoparticles on graphene. EG was ultrasonicated in 20 ml deionized water for 30 min to produce a graphene suspension. For NZF composition ($Ni_{0.5}Zn_{0.5}Fe_2O_4$), $NiCl_2.6H_2O$, $ZnCl_2$, $FeCl_2.4H_2O$ and $FeCl_3.6H_2O$ with molar ratio 1:1:2:2 were dissolved into 180 ml deionized water under nitrogen protection for 30 min. The graphene suspension was added to the solution of NZF at three different mass ratio of $FeCl_3$:EG (5:1,10:1 and 20:1) and stirred for a further 15 min. With vigorous stirring, NaOH solution was added to the mixture dropwise to maintain pH of 10-11. The reaction was carried out for 30 min. The suspension was then repeatedly washed

[a] Corresponding author : ing.kong@nottingham.edu.my

DOI: 10.1051/matecconf/201713601007

until pH ~7, filtered, dried at 60°C in air for 24 hours. In the absence of graphene, the NZF nanoparticles were obtained via the same procedure for comparison.

2.2 Characterization

X-ray diffraction (XRD) was conducted using a Bruker D8 Advance system based on reflectance operated at 45 kV and 35 mA with CuKα_1 radiation (λ = 1.54059 Å) in the 2θ range 10-80°, in steps of 0.04°. The morphology of the samples at different magnifications was observed by field emission scanning electron microscopy (FESEM) equipment Quanta400F (FEI), conducted in low vacuum mode, with an accelerating potential of 200 V-30 kV. The magnetic properties of the hybrids were studied using vibrating sample magnetometer (VSM), model Lakeshore 7404 Series at room temperatures (25 °C). The measurements were carried out in a maximum field of 12 kOe. Magnetic parameters such as saturation magnetization (M_S), remanence (M_R), and coercivity (H_C) were determined. For microwave absorbing properties, specimens were prepared by uniformly mixing 30 wt·% NZF and NZFG nanohybrids with paraffin wax at 70 °C and pressed into toroidal shape with inner diameter of 2.9 mm and outer diameter of 6.9 mm. The microwave absorption parameters of complex permittivity and complex permeability were measured at the frequency range 1-18 GHz by transmission/reflection method using an Anritsu 37247D microwave vector network analyzer with GPC-7 mm coaxial line.

3 Results and Discussion

The microstructure of NZF and NZFG nanohybrids were analysed by XRD technique and is shown in Fig. 1. All the diffraction patterns show broad peaks which indicate particles of fine crystallite sizes. NZF nanoparticles synthesized in the absence of graphene show characteristic peaks at 2θ = 30.18°, 35.43°, 43.11°, 52.15°, 56.90°, and 62.59° that can be assigned to (220), (311), (400), (422), (511) and (440) lattice planes of cubic unit cell of ferrite. These values match the single crystalline spinel form of the Ni-Zn ferrite (JCPDS Card No. 8-234). In all the Ni-Zn ferrite-graphene nanohybrids, the peaks of NZF appeared intense and are matched well, which confirms that NZF nanoparticles were successfully deposited on graphene. The diffraction peak at 2θ = 26.5° corresponds to the (002) plane of graphene. An additional peak which appears at 2θ = 64.4° (marked by *) shows the presence of α-Fe_2O_3 phase impurity. According to [11,12], this impurity exists when low amounts of precipitating agent were used. The average crystallite size based on Scherrer's formula (Eq. 1) were estimated to be 12.47 nm, 8.72 nm, 10.27 nm and 13.43 nm for NZF, NZF5G1, NZF10G1 and NZF20G1, respectively. The results show that the average crystallite size for NZFG nanohybrids increases with the mass ratio of $FeCl_3$:EG.

$$D = 0.89\lambda/\beta cos\theta \qquad (1)$$

where D is the crystallite size, λ is the X-ray wavelength, θ is the diffraction angle and β is the half width of the observed diffraction peak.

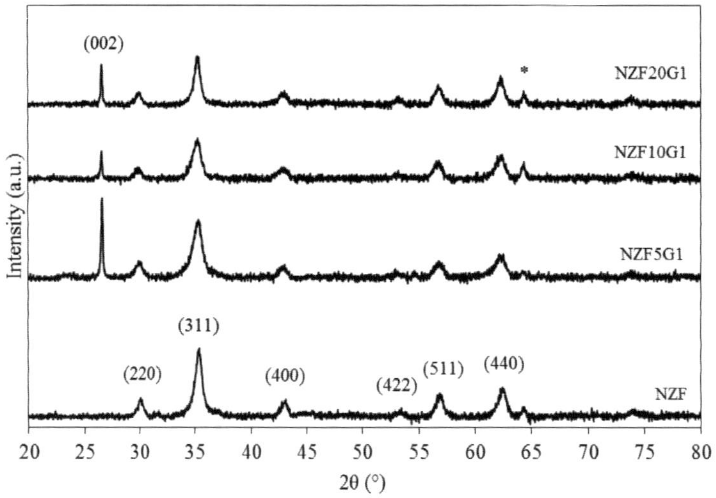

Figure 1. X-ray diffractograms of NZF, NZF5G1, NZF10G1 and NZF20G1

DOI: 10.1051/matecconf/201713601007

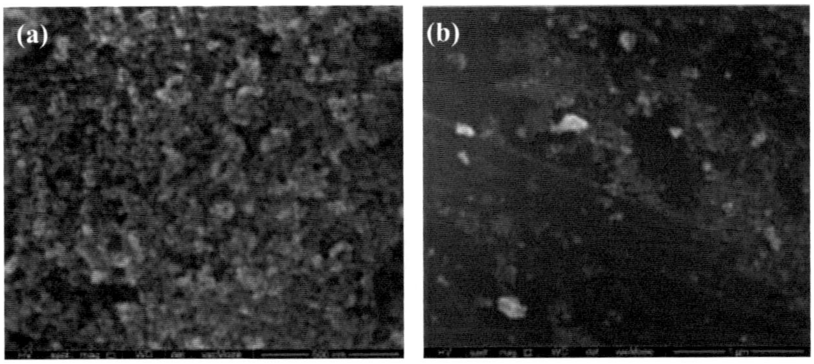

Figure 2. SEM images of (a) NZF and (b) NZF10G1.

Fig. 2 shows the SEM images of NZF and NZFG nanohybrids. It can be seen in Fig. 2(a) that the as-synthesized NZF nanoparticles have taken a spherical shape. As for NZFG nanohybrids, Fig. 2(b) shows that NZF nanoparticles were densely distributed over the surfaces of graphene sheets. The images indicate success in the formation of NZF in the presence of graphene.

The magnetization curves of NZF and NZFG nanohybrids are shown in Fig. 3. The values of magnetic measurements were determined from hysteresis loops and were summarized in Table 1. NZF nanoparticles recorded the highest M_S of 42.50 emu/g, which is higher than that produced in [5,13] but lower than that produced in [9]. A thin hysteresis was observed, showing the ferromagnetism of the nanoparticles at room temperature. As for NZFG, M_S values increase along with the mass ratio of $FeCl_3$ to EG. Due to the non-magnetic nature of graphene, the increment in M_S of the nanohybrids is likely driven by the amount and size effect of the NZF nanoparticles. In terms of size effect, larger sized nanoparticles tend to have higher M_S due to lesser surface distortions and magnetocrystalline anisotropy distortions of the nanoparticles [14]. Compared to M_S, relatively smaller M_R values for the samples implies that most of the spin moments has returned to their original state.

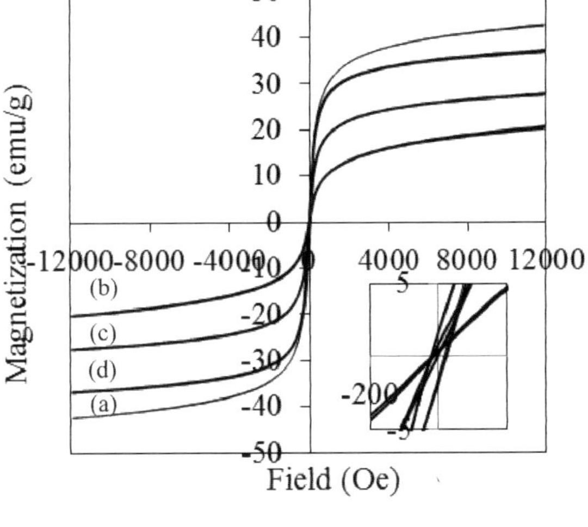

Figure 3. Magnetization curves of (a) NZF, (b) NZF5G1, (c) NZF10G1 and (d) NZF20G1.

Table 1. Magnetization characteristics of NZF, NZF5G1, NZF10G1 and NZF20G1.

Sample	M_S (emu/g)	M_R (emu/g)	H_C (Oe)
NZF	42.50	1.47	17.92
NZF5G1	20.46	0.25	11.14
NZF10G1	27.60	0.66	13.89
NZF20G1	36.90	1.51	19.19

The microwave absorption properties of wax composites containing 30 wt·% NZF and NZFG nanohybrids (NZF5G1, NZF10G1, NZF20G1) were determined by its complex permittivity ($\varepsilon_r = \varepsilon' - j\varepsilon''$) and complex permeability ($\mu_r = \mu' - j\mu''$), which can be measured using a vector network analyzer. Based on the measured microwave parameters, the reflection loss (R_L) can be calculated according to the transmission line theory using the two formulas below:

 DOI: 10.1051/matecconf/201713601007

$$R_L = 20 \log \left| \frac{(Z_{in} - Z_0)}{(Z_{in} + Z_0)} \right| \quad (2)$$

where R_L denotes the reflection loss in decibels (dB), Z_0 is the characteristic impedance of free space/air (377 Ω), Z_{in} is the input characteristic impedance of absorber. Z_{in} can be expressed as:

$$Z_{in} = Z_0 \left(\frac{\mu_r}{\varepsilon_r}\right)^{1/2} \tanh\left\{j\left(\frac{2\pi fd}{c}\right)(\mu_r \varepsilon_r)^{1/2}\right\} \quad (3)$$

where ε_r and μ_r are the measured relative complex permittivity and permeability, c is the velocity of light, f is the frequency of the microwave and d is the thickness of absorber.

Fig. 4 shows the calculated R_L of NZF and NZFG nanohybrids with a thickness of 7 mm. It can be seen that the R_L for NZF reaches a minimum of –31.56 dB at 7.33 GHz. As for NZFG nanohybrids, varying the NZF content varied the maximum absorption and position of the dips of the minimum R_L. First the minimum R_L increases from -30.65 dB at 7.15 GHz to -37.57 dB at 7.54 GHz and then decreases to -27.42 dB at 7.24 GHz. It is noteworthy that the frequency related to the R_L minimum lies within the C-band (4-8 GHz) of the electromagnetic spectrum, which is normally used for satellite communications. The bandwidth corresponding to R_L below –10 dB for NZF5G1, NZF10G1 and NZF20G1 are 8.00, 7.30 and 7.00 GHz, respectively. Based on the results, the enhancement in the microwave absorption properties for can be related to the synergistic effects of graphene coupled with NZF nanoparticles. When the dielectric and magnetic characteristics are matched well as in the case of NZF10G1, maximum absorption of the microwave energy occurs.

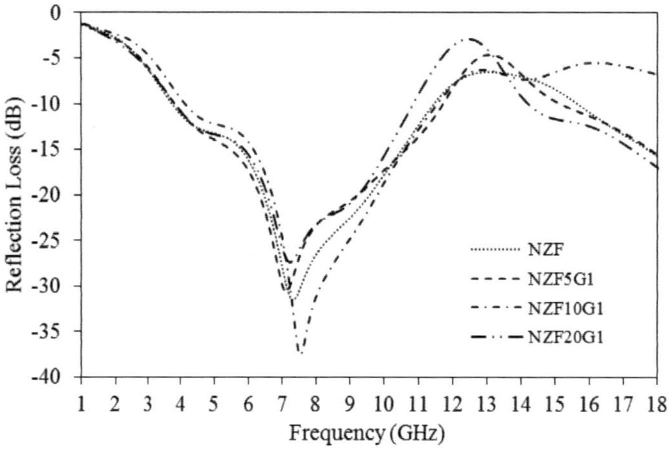

Figure 4. Frequency dependence of reflection loss for NZF and NZFG nanohybrids

4 Conclusion

In conclusion, a facile method was demonstrated to synthesis Ni-Zn ferrite-graphene (NZFG) nanohybrids. This method is capable of producing fine crystallite size nanoparticles with good crystallinity. The nanohybrids also showed good magnetic properties owing to a good distribution of NZF nanoparticles over the graphene sheets. Enhancement in the microwave absorption properties due to synergistic effect between NZF nanoparticles and graphene was seen when the NZF content in the nanohybrids is 79 wt·%. It is anticipated that these nanohybrids can be used as fillers in developing multifunctional polymer composites.

References

1. C. Wang, C. Feng, Y. Gao, X. Ma, Q. Wu, Z. Wang, Chem. Eng. J. 173, 92-97 (2011)
2. Y. Fu, Q. Chen, M. He, Y. Wan, X. Sun, H. Xia, et al., Ind. Eng. Chem. Res. 51, 11700-11709 (2012)
3. A.K. Geim, Science 324, 1530-1534 (2009)
4. I.H. Gul, W. Ahmed, A. Maqsood, J. Magn. Magn. Mater. 320, 270-275 (2008)
5. G.S. Shahane, A. Kumar, M. Arora, R.P. Pant, K. Lal, J. Magn. Magn. Mater. 322, 1015-1019 (2010)
6. A.S. Albuquerque, J.D. Ardisson, W.A.A. Macedo, M.C.M. Alves, J. Appl. Phys. 87, 4352 (2000)
7. S.M. Olhero, D. Soma, V.S. Amaral, T.W. Button, F.J. Alves, J.M.F. Ferreira, J. Eur. Ceram. Soc. 32, 2469–2476 (2012)
8. K. Nishimura, M. Abe, M. Inoue, IEEE Trans. Magn. 38, 3222–3224 (2002)
9. X. Lu, G. Liang, Q. Sun, C. Yang, Mater. Lett. 65, 674–676 (2011)
10. O. Perales Perez, K. Tohji, Y. Umetsu, J. Alloys Compd. 290, 129–136 (1999)
11. S. Thakur, S.C. Katyal, M. Singh, J. Magn. Magn. Mater. 321, 1–7 (2009)
12. K. Gandha, K. Elkins, N. Poudyal, J. Ping Liu, J. Appl. Phys. 117, 17A736 (2015)
13. B.P. Rao, A.M. Kumar, K.H. Rao, Y.L.N. Murthy, O.F. Caltun, I. Dumitru, et al., Adv. Mater. 8, 1703–1705 (2006)
14. C. Jiang, R. Liu, X. Shen, L. Zhu, F. Song, Powder Technol. 211, 90–94 (2011)

Fracture Analysis of CNG High Pressure Container using Fractography and Measurement of Property

Eui-Soo Kim[1,a]

[1]*Dept. of Safety Engineering, Korea National University of Transportation, Chungju-si, Chungbuk 27469, Korea*

Abstract. Bursting accidents of pressure containers due to design and manufacturing defects are frequently occurring. Due to high-pressure gas or harmful substances, when this vessel is fractured, it can lead to catastrophic disasters. Especially, in the event of bursting accident of composite pressure vessel for CNG bus, many unspecified people can be damaged. Most of the accidents were caused by problems in the manufacturing process. The manufacturing process for TYPE2 pressure vessel is very complicated such as three drawing processes, two ironing processes and one spinning process. In the middle of process, various heat treatments are performed for imparting toughness and removing residual stresses. It should cause a serious problem such as bursting and fragmentation of the pressure container due to defects of this process. In this research, the fracture cause of CNG vessel is evaluated through fractography and measuring material property using IIT and analysis of chemical composition.

1 Introduction

Bursting accidents of pressure vessels due to design and manufacturing defects are frequently occurring. Unlike other structures, Most of the high-pressure vessels contain high-pressure gas or toxic and harmful substances. When this vessel is bursting, it can lead to catastrophic disasters.[1] Especially, in the event of bursting accident of composite pressure vessel for CNG bus, many unspecified people can be damaged. Therefore, the safety and stability of CNG pressure vessel is very important. TYPE 2 container mainly used in Korea is reinforced with a circumferential direction by hoop wrapping glass fiber on steel liner. The CNG composite pressure vessel was made of a metal liner with epoxy-glass fiber composite material. Most of the fracture accidents in TYPE 2 were caused by problems in the manufacturing process. Especially, defect of the heat treatment process may cause a serious problem such as bursting and fragmentation of the pressure vessel. In this research, the investigation of explosive accident resulting from fracture of the pressure vessel is conducted by analyzing failure mechanism and any of the defect of the design and manufacturing process through fractography and measurement of mechanical and chemical property and comparison of stress state using computational numerical analysis in term of forensic engineering and suggest how to improve stability and reliability of CNG pressure vessel .[2][3]

2 Background of explosive accident

This accident was caused by bursting of CNG container loaded in the vehicle while charging the CNG bus at the CNG gas station as shown in Figure 1. The lower dome of the bursting pressure vessel is separated from the main body and this caused damage to the surrounding area by penetrating the upper part of the bus that was parked nearby. The accident CNG bus was equipped with a total of eight metal liners which is charged through connection pipes and the explosive liner is the sixth liner from the front of the bus.[4] The charger is equipped with three safety devices: mechanical safety valve, pressure switch, and pressure controller. The charging status of the charger is monitored and this record is stored automatically. Looking at the charging system, At the time of the accident, the gas pressure to the vessel is about 150 ~ 180 bar and malfunctions & specific defects are not observed. Therefore, possibility of accident due to charging overpressure and surrounding environment is very low. After all, the root cause of the explosive accident can be found in the problem of container itself such as defects of the design and manufacturing. To identify the exact causes of the CNG vessel failure, it is necessary to collect and investigate the fracture fragments of the vessel, which include composite parts, clamps.

Figure 1. Scene of explo sive accident of CNG buses

3 Fracture analysis of CNG container

3.1 Manufacture and charging process

When CNG is stored in the vessel by using a 4-stage reciprocating compressor, the gas inlet pressure of 3.5~4.5 bar can be increased by at least 200 bar. Compressed gas from the vessel is supplied to the inner container of the engine after reducing the high pressure of gas to a working pressure level through a pressure reducing valve. CNG composite pressure vessel for the purpose of containing high pressure gas have complicated manufacturing process. The pressure vessel of CNG vehicles is divided into four types which is dependent on the type of materials and composite materials used. TYPE 2 mainly used in CNG buses of Korea is reinforced with a circumferential direction by hoop wrapping glass fiber on steel liner. The TYPE 2 pressure vessel was made of a metal liner (34CrMo4 material) with epoxy-glass fiber composite material.[5] The composite material was wound around the cylinder part of the metal liner, forming a composite overwrapped pressure vessel. The manufacturing process of fractured metal liner consists of three drawing processes and two ironing processes and one spinning process for finalizing the dome part as show in Figure 2. In the middle of process, various heat treatments such as quenching, tempering, and annealing are performed for imparting toughness and removing residual stresses. The heat treatment process is the key process in the manufacturing of metal materials. Therefore, if there is a defect in the heat treatment process, it may cause a serious problem such as bursting and fragmentation of the pressure vessel.

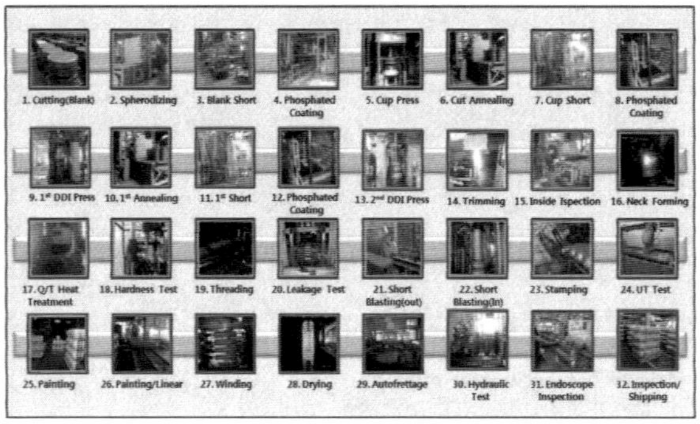

Figure 2. Manufacture process of CNG container

3.2 Visual inspection

The lower dome of the liner is fractured and separated from the body as shown in Figure 3. The traces of damage before an accident are not observed in the laminate of the composite material and the composite part was burst dependently by fracture of inner liner. In the result of observation after removing composite material, fracture of the liner was progressing along the axial direction. Also, severe corrosion patterns are observed on the surface around the fracture section and marks of the combustion reaction such as soot is not identified inside and outside the liner. A chevron mark which can determine the point of the crack initiation and the direction of the crack propagation in the axial fracture section of the lower dome and the body is observed. In the result of watching this mark, the shape of serration was observed in the section of the circumferential direction at a point 4 cm from the point where the lower dome of the liner meets the body. It is recognized that the crack propagates from the division where the shape of serration is observed to the axial direction. Looking at the fracture section of crack initiation, The corrosion layer is locally observed and the forms of brittle and ductile fracture are simultaneously identified.

Figure 3. Fracture and corrosion pattern of CNG container

3.3 Fractography using SEM and OM

The crack point on the crack initial surface was carefully observed by scanning electron microscopy (SEM) as shown in Figure 4., Figure 5. In case of TYPE 2 pressure container, metallic liner is manufactured of one circular blank by drawing and ironing continuous process. Depending on the process characteristics, vertical and parallel directions in the extrusion direction, the mechanical and structural properties in the container are different. Therefore, in order to examination mechanical properties and micro-structure of crack point, Test is examined according to the extrusion vertical direction (one direction) and parallel direction (second direction). As a result of the inclusion inspection, special point was not identified in all directions as shown in Figure 6. Martensite structure was observed as a result of micro-structure examination. However, it shows fine martensite structure in the cross section perpendicular to the extrusion direction, but coarse martensite was identified in the cross section parallel to the extrusion direction as shown in Figure 7. Martensite is observed in the entire region of the vessel which must have the quenching and tempering process, and it is judged that there is a defect in the heat treatment of the tempering skip. It can be seen that it is very brittle and the fragmented because the toughness is not given. As a result of FE-SEM observation, we know that the precipitation was seen much more in vertical cross section than the parallel cross section of the extrusion direction.

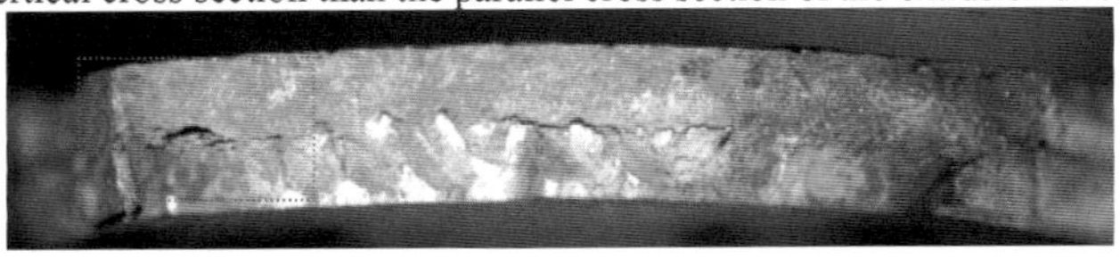

Figure 4. Microscope mage of crack initial area

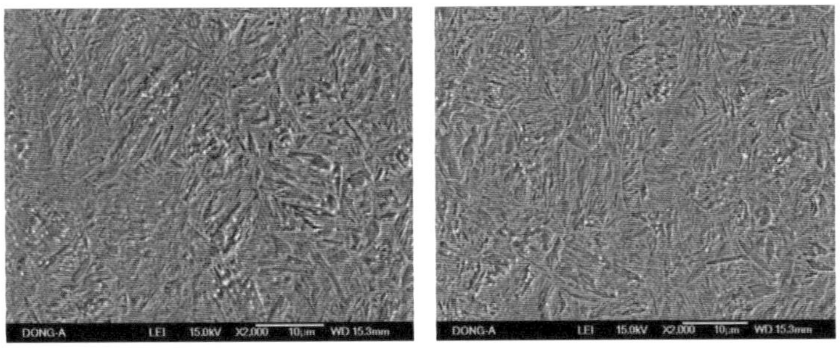

(a)Vertical to extruding direction (b) Horizontal to extruding direction

Figure 5. SEM Images of crack initial point (2000x)

(a)Vertical to extruding direction (b) Horizontal to extruding direction

Figure 6. OM Images for inclusion(100x)

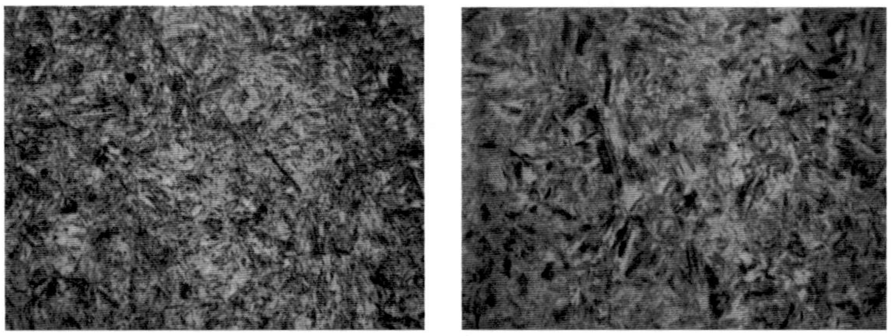

(a)Vertical to extruding direction (b) Horizontal to extruding direction

Figure 7. OM Images for martensite (1000x)

3.4 Measurement of property

In order to ascertain the singularity of the chemical composition of the liner part of the fracture initiation point, as a result of component analysis according to photoelectric photometric emission spectroscopic analysis method of iron and steel (KSD 1652: 2007) as shown in Table 1.[6] The value is satisfying a given design condition. In order to evaluate mechanical properties, specimens were collected for each site along each crack propagation line. Mechanical properties were evaluated by IIT(Instrumented Indentation). Based on the evaluation process of ISO TR29381and KS B 0950:2002, IIT is designed to determine the mechanical properties by putting an indentation in the specimen's surface.[7][8] This method is known to be a powerful tool to probe the mechanical response of various engineering materials when irregularly shaped fragments of mechanical parts need to be analyzed. The results are shown in Table 2. comparing with yield strength (840 MPa) and longitude value (270 to 321 HV) of the design condition. All of the value far exceeded the design condition. Here, n is the work hardening index, YS is the yield strength, UTS is the breaking strength, E is the elastic modulus, HV is the Vickers hardness, and HR - A is the Rockwell hardness. As for material analysis, the liner satisfies the chemical composition ratio, hardness value does not satisfy the design range. Generally, the variation in the hardness value results from inappropriate heat treatment such as quenching and tempering in the manufacturing process. In the case of the container of this accident, the value of hardness far exceeds the design condition by inappropriate heat. If the value of hardness increases, brittleness becomes high and it induces a condition which is extremely vulnerable to hydrogen brittleness and fatigue failure.

Table 1. Result of chemical composition

Copo.	C	Si	Mn	P	S	Cr	Ni	Mo
Standard	0.32-0.37	0.15-0.35	0.60-0.90	~0.015	~0.010	0.90-1.15	~0.25	0.20-0.25
Measure	0.345	0.222	0.73	0.006	0.003	0.985	0.246	0.188

Table 2. Result of mechanical property

E. V	n	YS(MPa)	UTS(MPa)	E(Gpa)	HV	HR-A
Mean of val.	0.201	1043.2	1466.3	208.3	490.1	74.9

4 Conclusions and recommendations

In this research, in order to investigate the cause of the explosion accident of the CNG pressure vessel, the fracture evaluation of the container accomplish using fractography and measurement method of mechanical engineering. As a result of analysis, A local corrosion layer and pattern of cleavage fracture is identified. In term of the material analysis, although the liner is satisfied with the chemical composition ratio, the hardness value of the mechanical properties does not satisfy the design value. Because brittleness is high due to improper heat treatment, that increase a possibility of causing cleavage fracture with little plastic deformation. We recommend how to improve the safety and reliability of pressure vessels. Firstly, we have to prevent non-experts from making an error by automation everything from design to inspection. Secondly, We suggest to replace the TYPE 2 pressure vessel with TYPE 3 pressure vessel which is more effective than the TYPE 2 in term of LBB(Leaked Before Break) effect. Thirdly, We ask for increasing supply of the ground bus which mount pressure vessel to the upper of bus in order to release immediately inner gas without any residue when CNG leaked.

Acknowledgement

This research was supported by Basic Science Research Program through the National Research Foundation of Korea(NRF) funded by the Ministry of Education(2017R1D1A1B03028163)

References

1. E.S. Kim, N.K. Park, Y.H. Kim, KOSOS, *A forensic engineering study on bursting accident of composite pressure vessel in CNG bus*, **23**, 15-21 (2008)
2. D.H. Michael, B.E. Dale, R.S. Anand, Fractography in Failure Analysis of Polymers, *Fractography as a Failure Analysis Tool*, 6-22 (2015)
3. S.M. Goushegir, J.F. Dos Santos, S.T. Amancio-Filho, Composites Part A: Applied Science and Manufacturing, *Failure and fracture micro-mechanisms in metal-composite single lap joints produced by welding-based joining techniques*, **81**, 121-128 (2016)
4. S. Chamberlain, M. Modarres, Risk Analysis, *Compressed Natural Gas Bus Safety: A Quantitative Risk Assessment*, **25** (2005)
5. E.S. Kim, J.H. Kim, Y.S. Park, C.J. Choi, KSPE, *Development of an automated design system of CNG composite vessel using steel liner manufactured by D.D.I process*, **20**, 205-213 (2003)
6. Korean Standards, *Iron and steel —Method for spark discharge atomic emission spectrometric analysis*, KS D **1652**:2016, (2016)
7. K.H. Kim, Y. Choi, S.H. Kim, J.S. Lee, D.I. Kwon, Spring conference of KSME, *Application of Instrumented Indentation Technique for Nondestructive/Mechanical Properties Measurement of In-service Components and Materials*, 336-343 (2005)
8. A. Sina, C. Xi, M.S. Andrea, G. Marco, International Journal of Mechanical Sciences, *Investigation of the mechanical behaviour of lithium-ion batteries by an indentation technique*, **105**, 1-10 (2016)

 DOI: 10.1051/matecconf/201713601009

Preparation and Sensitivity Measurements of Graphene Oxide-RDX Composite

Mahmoud Y. Zorainy[1], Waleed F. Aly[1], Ahmed Elbeih[1,a], and Mohamed Gobara[1]

[1]*Military Technical College, Kobry Elkobbah, Cairo, Egypt*

Abstract. The effect of graphene oxide (GO) on the safety characteristics of 1,3,5-trinitro-1,3,5-triazinane (RDX) was studied in this work. Graphene oxide was prepared and was investigated to form a composite based on GO-RDX by solvent-antisolvent slurry technique. For comparison, different polymer bonded explosives (PBXs) based on RDX bonded by viton A, fluorel or polymethyl-methacrylate binders were studied and designed as RDX-Viton, RDX-Fluorel and RDX-PMMA respectively. Sensitivities to impact and friction of the presented samples as well as the pure RDX was measured. The ignition temperatures were determined and the activation energies were calculated by using the ignition delay method. Results of x-ray diffraction and scanning electron microscope proved that RDX crystals were coated by a thin layer of GO. The impact sensitivity of GO-RDX composite is lower than that of the other studied samples while the friction sensitivity is slightly higher. The ignition temperature of GO-RDX was lower than the other studied samples which indicates that the GO caused accumulation of the decomposition gaseous products and accelerated the decomposition process of RDX. GO is an interesting candidate material to be used for coating the explosive crystals instead of the polymeric matrices.

1 Introduction

In the field of energetic materials, it is well known that the advanced explosives have high performance and in the same time high sensitivities [1-3]. The researchers are seeking to obtain modern explosives with high performance and high safety during handling and transportations [4–6]. The sensitivities of the explosives could be enhanced by controlling the crystal shapes and sizes [7, 8]. Traditionally, the preparation of polymer bonded explosives (PBXs) was the optimum solution to increase the safety of the explosives without affecting their performance [9, 10]. Many publications reported the preparation methods and the characteristics of the prepared PBXs [11, 12]. The percentage of the polymer used to coat the explosive crystals has significant effect of the performance of the prepared PBXs [13]. The thermal behavior and decomposition kinetics of PBXs are important to produce and long life PBXs [14, 15]. 1,3,5-trinitro-1,3,5-triazinane (RDX) is an interesting explosive which has several applications if the field of energetic materials [2,3]. The sensitivity of RDX is consider high and cannot be pressed as a pure materials. Graphene oxide (GO) is a two-dimensional (2D) material and can be obtained by oxidation of graphite powders [16]. GO has been widely applied in material sciences due to its large theoretical specific surface area and abundant oxygen-containing groups, which can significantly alter the properties of materials it is combined with [17, 18].

In this work, a method for preparation of GO-RDX composite is presented. The effect of coating the explosive crystals by 2 wt% of GO on the impact and friction sensitivities was discussed, in addition to its effect of the activation energy and the ignition temperature of the pure RDX. For comparison, selected PBXs have been studied.

2 Preparation of graphene oxide (GO)

GO was prepared by improved Hummers method with some modification. Preparation of GO was prepared by improved Hummers method, four steps has been followed to apply this method (30). 1.st step putting 90 mL H_2SO_4 and 60 mL H_3PO_4 in beaker 250 ml, adding 3.0 g Graphite in this beaker with stirring in ice bath for 30 min, till the temp. Reach 0 °C, adding slowly 20 g $KMnO_4$, and stirring for 1 h. The second step was removing the beaker from ice bath and putting it hotplate at 60 °C for 24 hr. then adding a 500 ml of cold deionized water to the forms avoid temperature increase > 50 °C, adding 30 ml H_2O_2 the color turn from brown to yellow, adding 500 ml DW for 24 hr. the last step based on removal of the water superannuated adding 500 ml DW put the baker in ultrasonic bath for 1 hr at 40 GHz and wash by HCL 5% and DW using centrifuge at 10000 rpm until pH reach 6 and dry the precipitation by vacuum oven at 60 °C for 12 hr.

3 Preparation of GO-RDX composites

[a] Corresponding author : elbeih.czech@gmail.com

DOI: 10.1051/matecconf/201713601009

The formation of GO-RDX composites was prepared In brief; in a beaker of 250 ml 10 mg of the prepared GO was dispersed in 50 ml DMF and sonicated for 60 min. Then, 0.5 g of RDX was dissolved in 50 mL DMF with magnetic stirring and added to the GO suspension and assisted ultrasonic vibration, allowed for stirring for 2 h at 40°C. After RDX and GO was completely dissolved in DMF solvent, the mixture was dropped into the anti-solvent coldwater under magnetic stirring. The product was separated by centrifugation at 5000 rpm and washing by methanol and drying in vacuum oven at 50°C for 12 hr. The GO-RDX composite was prepared in with a theoretical mass fraction of GO of 2%. RDX was recrystalline in a similar process without addition of GO.

4 X-ray diffraction

The XRD patterns of GO, RDX, and GO-RDX by using a fully computerized X-ray diffractometer, Shimadzu XRD-6000 with Cu radiation λ=1.54056 A° was carried out at National Center for Radiation Research and Technology, Egyptian Atomic Energy Authority. The X-ray tube was operated at 40kV and 30mA anode current throughout the measurements. The above operation conditions were maintained during all the relevant measurements. The 2θ scan range from 4° to 90° continuous scanning with scan speed 8 (deg/min) and preset time 0.15 sec was used for the entire 2θ-range.

Many parameters can be concluded from X-ray analysis such as crystal structure, atomic arrangement, phase orientation, lattice parameters and grain size. The samples of pure GO, RDX and as-prepared GO-RDX composites were examined by XRD. As shown in Fig. (1), a sharp peak at 2 theta = 10.2o is related to GO powder with d-spacing = 8.0427 Å. Moreover, d-spacing value was increased due to the insertion of function groups between graphite layers. This indicated the high oxidation applied to graphite during graphene oxide preparation. These results highly agree with the previous literature of preparation GO [18]. The major diffraction pattern of RDX is almost the same as that of GO-RDX composites (such as 12.8°, 17.14° and 28.9°) indicated that interaction between GO and RDX.

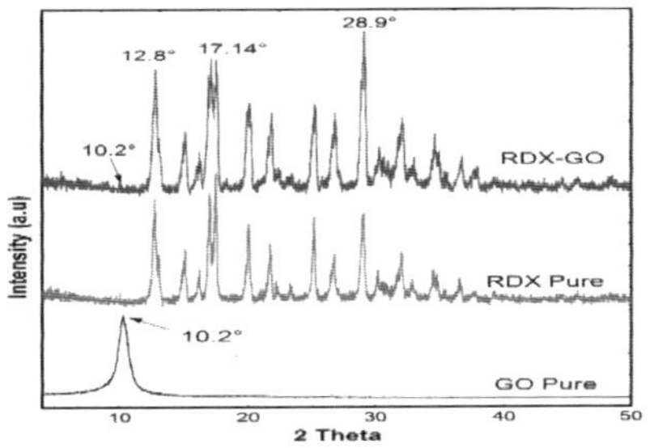

Figure 1. XRD of the studied materials.

5 Crystal morphology study

Scanning electron microscope (SEM) model Inspect S obtained from FEI company was used to study the crystal morphology of RDX and GO-RDX crystals. Figure 1 presents the SEM photos of the pure RDX crystals with magnification.

(a) (b)

Figure 2. SEM photos of RDX crystals.

(a) (b)

Figure 3. SEM photos of the GO-RDX crystals.

As shown in Fig. 2, pure RDX crystals are in the microscale and the crystals have particle size less than 100 μm. The RDX crystals have non-uniform shapes and there are cracks on the surface of the crystals. Also the surface of the crystals are rough and that is the reason of the high sensitivity of the pure RDX. The morphology of the prepared GO-RDX is shown in fig. 3, where the crystal have nearly the same size as the pure RDX. It means that the coating of RDX by GO does not affect the size of the used RDX. The crystals of the prepared GO-RDX have a smooth surface where the cracks of the pure RDX disappeared due to covering of the crystals by GO and it confirms that GO are precipitated on the surface of RDX crystals. Also the GO caused decrease to the sharp edges of the RDX crystals, as a result it could decrease the sensitivity of the pure RDX due to the modification of the morphology of the crystals. It is also clear that addition of 2 wt % of GO affect the morphology of RDX crystals.

6 Impact sensitivity measurements

The impact sensitivity of the prepared GO-RDX in addition to the studied PBXs based on RDX was determined by a standard impact tester with exchangeable anvil (Julius Peters [19]) where a 2 kg drop hammer was use. The tested samples had a volume of 50 mm^3. The results of the Probit analysis technique [20] was used to determine the probability of initiation of the tested samples. The sensitivity to impact was reported as the impact energy required for 50% probability of initiation of the samples and is reported in Table 1.

7 Friction sensitivity measurements

The friction sensitivity was obtained by using BAM friction test apparatus with applying the standard test conditions [19]. 0.01g of the studied explosive was spreaded on the surface of the porcelain plate and different loads were used to change the normal force between the porcelain pistil and the plate. Detection of the sample initiation was observed through sound, smoke appearance, or through smell of the decomposition products. The Probit analysis was used to determine the probability of initiation of the samples. However, only 50% of initiations was reported as the normal friction force required for initiation and reported in Table 1 as the friction sensitivity.

Table 1. The sensitivity measurements of the studied explosives.

No.	Explosive type	Impact sensitivity (J)	Friction sensitivity (N)	Ignition temperature (°C)
1	RDX [21]	5.6	120	215
2	GO-RDX	13.4	268	213
3	RDX-viton [10]	10.6	326	216
4	RDX-fluorel [10]	10.8	352	217
5	RDX-PMMA [10]	11.9	240	212

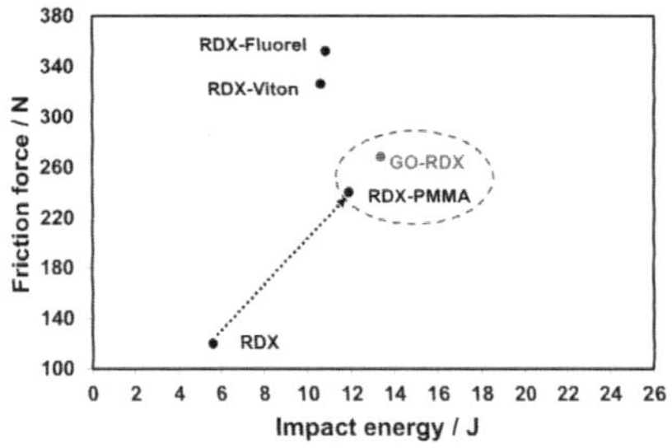

Figure 4. Comparison between the impact and friction sensitivities of all the studied samples.

The results of the impact and friction sensitivities are presented on fig. 4. It is clear that RDX has a high impact sensitivity while GO-RDX has the lowest impact sensitivity of all the studied samples (13.4 J) even that GO-RDX contains 98 wt% of RDX while the other PBXs have 9 wt% of RDX. It means that GO has a great influence on decreasing the impact sensitivity of RDX. This result might be due to the effect of the GO sheets which can dissipate the impact energy and decrease the effect of the potential energy on the RDX crystals and prevent the hot spot formation. On the other side, the friction sensitivity of GO-RDX is higher than RDX-Viton and RDX-Fluorel. This might be due to the characteristics of the fluorinated binders which has great effect on decreasing the sliding friction on the explosive itself. While GO-RDX has lower friction sensitivity than RDX-PMMA. It means that the results of GO-RDX were the optimum compared with the studied PBXs.

 DOI: 10.1051/matecconf/201713601009

8 Ignition temperature and ignition delay measurements

Chilworth deflagration test apparatus was used to determine the ignition temperature and ignition delay of the samples. Three test tubes contained 0.1 g from each sample were placed vertically into heating device at the same moment where the temperature was increased by fixed heating rate of 5 °C min^{-1}. Once the sample was ignited, the ignition temperature was recorded where the mean value was recorded in table 1. Ignition delay was obtained by increasing the temperature of the heating block over the ignition temperature and determine the delay time required for ignition of the samples at different temperatures. The time required for ignition of each sample is presented in table 2 at different isothermal temperatures. The kinetic parameters can be obtained by using the measurement of the ignition delay method as reported in literature [22]. Arrhenius equation was used to determine the activation energy of the studied samples by plotting a linear relationship between ln (time to ignition, t_{ig},) and 1000/temperature, T/K, of each sample as shown in Fig. 5. The activation energies of the samples were obtained from the slope of the curve and recorded in table 2.

Table 2. Ignition delay time (t_{id}) and activation energy (E_a) of the studied samples.

Sample	t_{id} / s at temperatures / ^{0}C								Ea / kJ mol^{-1}	R^2
GO-RDX	250 33	260 21	270 10	280 6	290 4	300 2	310 -	320 -	131.3	0.9918
RDX-Fluorel	250 39	260 25	270 16	280 10	290 8	300 7	310 6	320 5	74.8	0.9549
RDX-Viton	250 41	260 26	270 17	280 11	290 9	300 7	310 6	320 4	81.5	0.9832
RDX-PMMA	250 34	260 23	270 13	280 7	290 6	300 5	310 4	320 2	96.2	0.9649

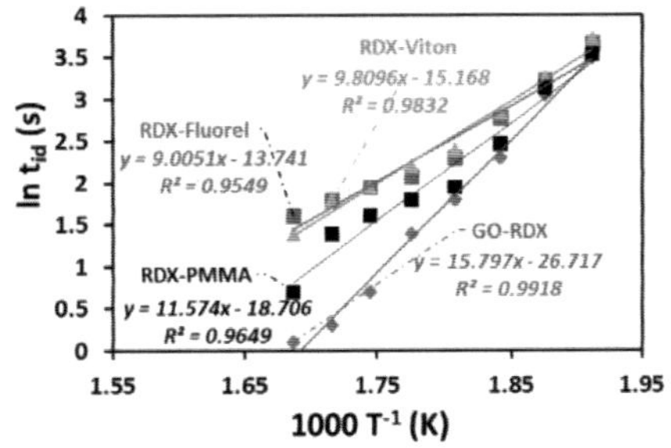

Figure 5. A linear relationship between ln (tig) and 1000 T-1 for all the studied samples.

From fig. 5, it is clear that GO-RDX has the highest activation energy of all the studied samples while RDX-fluorel and RDX-viton have very close results. It means that the increase of the activation energy of GO-RDX might be due to the interaction forces between the graphene oxide with the crystals of RDX where the decomposition is difficult. Also the fluorinated binders (viton and fluorel) have almost the same effect on the pure RDX.

9 Conclusion

From this study it was concluded that the method used for preparing composite of GO-RDX produced crystals with regular shape and smooth surface which enhanced the sensitivity characteristics of GO-RDX. A composite of RDX with 2 wt% of GO has sensitivity to impact lower than PBXs containing 9 wt % binder. GO-RDX has friction sensitivity in the same level of RDX-PMMA and higher than RDX-viton. These results clarified that GO has obvious effect on decreasing the sensitivity due to uniaxial compression while it has lower effect on shear slide with a fixed volume (sensitivity to friction). The activation energy of GO-RDX obtained by the ignition delay is higher than the other studied PBXs which means that the interaction between GO and RDX was strong and required a higher activation energy than the traditional PBXs. GO is a candidate to replace the binder systems which are used to decrease the sensitivity of explosives.

References

1. A. Elbeih, J. Pachman, S. Zeman, W.A. Trzcinski, M. Suceska, Propellants, Explosives, Pyrotechnics, **38**, 238 (2013).
2. H.-H. Licht, Propellants Explosives Pyrotechnics, **25**, 126 (2000).
3. S. Zeman, A. Elbeih, Z. Akštein, Hanneng Cailiao/Chinese Journal of Energetic Materials, **19**, 8 (2011).
4. Q.-L. Yan, S. Zeman, P.E. Sánchez Jiménez, T.-L. Zhang, L.A. Pérez-Maqueda, A. Elbeih, Journal of Physical Chemistry C, **118**, 22881 (2014).
5. A. Elbeih, S. Zeman, M. Jungova, Z. Akstein, Central European Journal of Energetic Materials, **9**, 131 (2012).
6. T.M. Klapötke, *Chemistry of high-energy materials*, (Walter de Gruyter GmbH & Co KG, 2015).

7. A. Elbeih, A. Husarova, S. Zeman, Central European Journal of Energetic Materials **8**, 178 (2011).
8. Q.-L. Yan, S. Zeman, A. Elbeih, A. Zbynek, Central European Journal of Energetic Materials, **10**, 509 (2013).
9. A. Singh, M. Kumar, P. Soni, M. Singh, A. Srivastava, Defence Science Journal, **63**, 622 (2013).
10. A. Elbeih, S. Zeman, M. Jungova, P. Vávra, Z. Akstein, Propellants, Explosives, Pyrotechnics, **37**, 676 (2012).
11. Q.-L. Yan, S. Zeman, R. Svoboda, A. Elbeih, Thermochimica acta, **547**, 150 (2012).
12. Q.-L. Yan, S. Zeman, F.-Q. Zhao, A. Elbeih, Thermochimica acta, **556**, 6 (2013).
13. A. Elbeih, M. Mohamed, T. Wafy, Propellants, Explosives, Pyrotechnics, **41**, 1044 (2016).
14. M. Abd-Elghany, T. Klapötke, A. Elbeih, S. Zeman, Journal of Analytical and Applied Pyrolysis, **126**, 267 (2017).
15. A. Elbeih, S. Zeman, J. Pachman, Central European Journal of Energetic Materials, **10**, 339 (2013).
16. C. Zhang, X. Cao, B. Xiang, J. Phys. Chem. C, **114**, 22684 (2010).
17. Z. Li, Y. Wang, Y. Zhang, L. Liu, S. Zhang, RSC Advances, **5**, 98925 (2015).
18. J. Wang, B. Ye, C. An, B. Wu, H. Li, Y. Wei, J. Energ. Mater, **34**, 235 (2016).
19. M. Suceska, *Testing Methods of Explosives* (Springer, Heidelberg, 1995).
20. D. Finney, *Probit Analysis* (Cambridge University, third edition, 1971).
21. A. Elbeih, T. Wafy, T. Elshenawy, Central European Journal of Energetic Materials, **14**, 77 (2017).
22. A. Elbeih, M. Abd-Elhgany, T. Elshenawy, Acta Astronautica, **132**, 124 (2017).

 DOI: 10.1051/matecconf/201713601010

Preparation and Characterization of Modified Silica-Epoxy Hybrid Ceramic Coatings

Mahmoud Y. Zorainy[1,a] Ahmed Nashaat[1], Yasser El-Shaer[1], Mohamed Gobara[1] and Ahmed Elbeih[1]

[1]*Military Technical College, Kobry Elkobbah, Cairo, Egypt*

Abstract. Hybrid sol-gel processing of inorganic-organic nanocomposites has been of a great interest over the last decades for being advantageous compared to the conventional addition methods of nanoparticles. In this study, a three-component system was adopted experiencing the design and preparation of different hybrid ceramic coatings based on Diglycidyl ether of Bisphenol A (Epoxy) and 3-Glycidyloxypropyl trimethyloxysilane (GLYMO) by sol-gel technique. The obtained hybrid coatings were cured using different hardeners, Diethylene triamine (DETA) as an organic linker and 3-Aminopropyl triethoxysilane (APTES) as an inorganic/organic linker. Microstructure assessment and the morphology of the prepared hybrids was investigated using FTIR and scanning electron microscopy (SEM) respectively. Mechanical properties (adhesion, and hardness) were determined. The degree of hydrophilicity of the hybrids was assigned depending on the contact angle measurements. Moreover, the thermal properties were investigated using thermogravimetric analysis (TGA). The results showed that the silica content plays an important role in determining the morphology as well as the mechanical, physical, and thermal properties of the coatings. The results showed an improvement in most of the properties of the hybrid coatings with increasing the silica content up to a certain extent. However, further increase in the silica content, leads to a clear deterioration.

1 Introduction

Sol-gel method is a low temperature processing technique that is used for the synthesis of the ceramic materials. A hybrid organic/inorganic has been thought of in order to achieve a homogeneous combination of both materials or somehow control their properties. In particular, silicon based hybrid materials using sol-gel processing is one of the major categories. The reason for the special role of silicon is the stability of the Si-C bond of the coupling agents used during the formation of the silica that in turns, contributes in the formation of these hybrid networks [1]. In general, coatings have developed side by side with ceramics until a crossroad was once noticed. Ceramic coating has been of great interest, where a specific material with certain properties is being selected to adhere to a surface (substrate) showing the desired properties of ceramics or a synergistic character of both ceramics and the substrate. Hybrid coatings compromise two components; an organic one that is responsible for toughness, adhesion and elasticity. While the other is inorganic, that characterizes some other properties like hardness and wear resistance. Silica sol–gel coatings form van der Waals bonds with the metallic substrate that transforms into a stable covalent bond during the curing process. Epoxy–silica coatings were studied by a number of researchers. Generally, epoxy resins have superior chemical and corrosion resistance, high adhesion, and curability at ambient temperature. However, they suffer from poor thermal and weathering stabilities, reduced mechanical properties and unsuitable cutting properties. This limits the epoxy resins in high-performance applications [2]. Using silica inorganic precursor within the epoxy resins to form epoxy–silica hybrid materials may improve these limits [3]. In this study, silane-modified epoxy resins from pre-hydrolyzed GLYMO was consequently developed in the presence of a relative concentration of 3-aminopropyl triethoxysilane (APTES) and diethylene triamine (DETA) as curing agents. Therefore, hybrid silica sol-gel coatings were prepared with different silica content. The coatings were applied on AA2024 then the effect of the silica content on some properties were investigated.

2 Materials and methods

2.1 Materials

As the organic phase precursor, DEGBA epoxy resin of EEW=450-500 g/equiv. was obtained from HUNTSMAN (China). The inorganic precursors, GLYMO and APTES were obtained from Sigma-Aldrich, USA. DETA as a curing agent was obtained from CDH, India. Acetone (solvent) as well as HCl (37 wt% in water) was purchased from BDH, England. All reagents were used as received without any further modification or processing. All coatings were performed at ambient temperature on freshly prepared coupons of Aluminium alloy 2024-T3 of dimensions (2.54×10.16×0.156) cm^3 obtained from Q-panel, USA. Coupons were

[a] Corresponding author : m.yossry@mtc.edu.eg

 DOI: 10.1051/matecconf/201713601010

cleaned with deionized water followed by ultrasonic cleaning in acetone and then dried for 30 min at 70 °C then stored in desiccator till their use.

2.2 Preparation of silica/epoxy sol-gel coating

The sol-gel technique depends on a series of hydrolysis and condensation reactions as shown in equations (1-3), as follows:

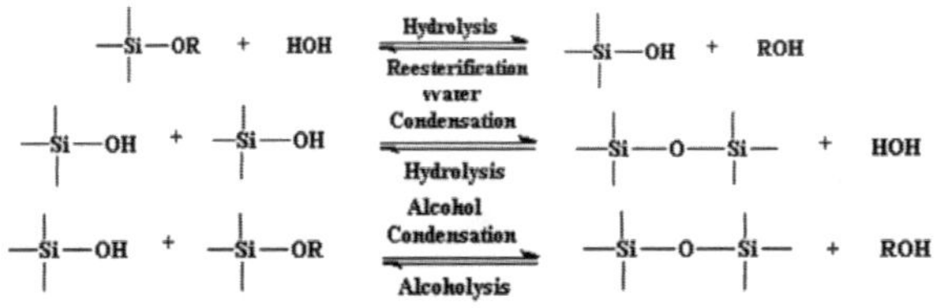

(1)

(2)

(3)

A mixture consisting of epoxy resin and GLYMO in the presence of acetone was prepared according to the wt. ratio of 1:1:6 respectively. The mixture was stirred for 15 min. till the complete dissolution of the epoxy resin. Conc. HCl was added to the solution preserving a water ratio of 3 with respect to the amount of GLYMO, keeping the pH within the range of 4.5±0.3. The solution was stirred for 1 hr. while heating to 100oC then the solution was left overnight. The solution then divided into six portions in which different amounts of hardeners (DETA and APTES) were to be added according to Table 1. Each portion is to be stirred for 30 min. then spayed over the Al coupons. The thickness of the coat is about 50±5μm.

2.3 Characterization and measurements

FTIR analysis was performed using a JASCO 4100 FTIR spectrometer. The surface morphologies of the used coupons were investigated using SEM (Zeiss EVO-10). The contact angle was measured using a ramé-hart goniometer model 250 adopting the sessile drop model with a syringe-out technique. (TGA) studies on the samples were executed on SHIMADZU TGA-50H Analyser. The thickness of the prepared hybrid coat sprayed over the aluminium substrate was to be measured by the coat thickness gauge Kett LZ-330C. The adhesion measurements were performed on double-lap joints with the coat prepared to be used as the adhesive material. A pull-off adhesion test was performed according to ASTM 5656 at room temperature on a GALDABINI QUASAR 100. The Vickers hardness measurements were performed on INSTRON TUKON 2100B.

3 Results and discussion

Table 1. Different compositions of the prepared hybrids with the corresponding mechanical and physical results.

Sample	GLYMO:Epoxy ratio*	Hardeners (APTES:DETA) ratio	Shear stress (MPa)	Hardness		Contact angle	Silica Content (%)
				HV	Mean diagonal length (μm)		
A0	1 : 1	0 : 1	1.13	147.3	354.8	69.7°	12.12%
A1	1 : 1	1 : 4	1.67	188.8	313.4	73.6°	15.04%
A2	1 : 1	1 : 1.5	1.87	221.0	289.7	74.9°	18.31%
A3	1 : 1	1 : 0.67	1.29	170.8	329.5	70.5°	21.36%
A4	1 : 1	1 : 0.25	1.20	149.1	352.7	62.5°	24.12%
A5	1 : 1	1 : 0	0.92	134.4	371.5	59.6°	27.59%

*wt. Ratio

3.1 FTIR Analysis

Using the Fourier Transform Infrared technique to investigate the preparation of the hybrid network, the spectra of the main precursors of the hybrid coat were studied individually as shown in Figure 1.

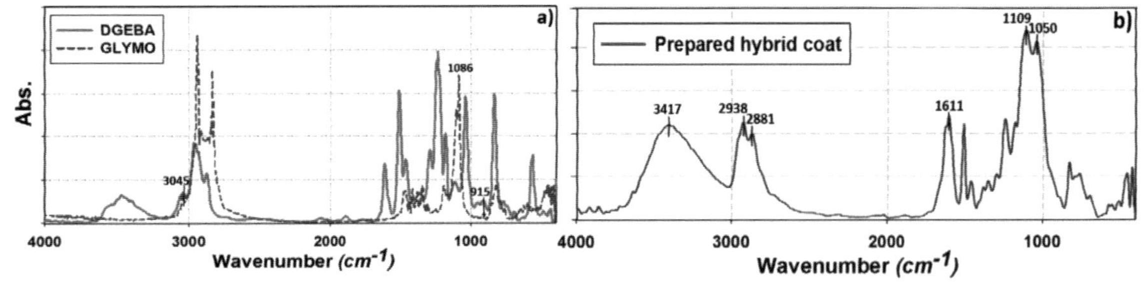

Figure 1. FTIR spectra of a) DGEBA and GLYMO, b) The prepared hybrid coat.

Primarily, the spectrum of each precursor was investigated and the main structural bands were focused upon as shown in Figure 1.a. Starting with DGEBA (Epoxy), the peaks aiming at both the benzene and the oxirane rings (found around 3045 and 915 cm^{-1}) were specified. For the main ceramic-constituent precursor, GLYMO spectrum was clarified for the Si-O-C bands (at 1200-1000 cm^{-1}) and the oxirane ring peaks [4-6]. The spectra of the prepared hybrid coats were compared to each other and it was found to be approximately identical. Therefore, a selected spectrum of sample A2 was shown in Figure 1.b as a representative for all hybrid coats. The spectrum was revised for the oxirane ring peaks at (3047 and 916 cm^{-1}) that have almost diminished. Also, the bands related to the amino groups along with the hydroxyl groups at the ranges of (3570-3200 cm^{-1}) and (1620-1550 cm^{-1}) were highly recorded, indicating the breakage of the oxirane ring and the existence of uncondensed hydroxyl groups with a residual amount of the curing agent. Aside from the oxirane ring, the Si-O-C bands were studied. Once the silica network are created, the sharp peaks are replaced with those broad peaks as found in Figure 1.b at (1050 and 1109 cm^{-1}). Conservatively, a peak at the range of 3420 cm^{-1} is taking place form the hydrolysed Si-OH species. Also, the peak at 3420 cm^{-1} may indicates the residuals of the corresponding released alcohol [7, 8]. Therefore, the silica network formation through the sol-gel technique is successfully achieved.

3.2 Mechanical characterization

3.2.1 Adhesion test

Applying different ratios of APTES to DETA to the (Epoxy-GLYMO) base coat, in which APTES is used as another source of silica, has made the synthesis of a silica based hybrid coat more integrated. Appling the test for the prepared combinations, the load-deformation curves were analyzed and the max. load applied as an indication for adhesion strength were assigned. As a result, the shear stress for a double area of (2.5 x 7 cm^2) was concluded as shown in Table 1. As shown, introducing APTES for samples A1 and A2 offered a higher adhesion than that of the DETA only (A0). Then, combinations of higher contribution of APTES to DETA (A3 and A4) have shown a gradual decrease reaching the minimum at APTES only (A5). At first, the addition of APTES produces a more homogeneous, compatible hybrid coat that gradually increases adhesion. A more compatible coat would offer a contribution of the hydrolyzed silanol groups to the bonding with the substrate forming a Si-O-Al bond increasing adhesion. Further increase of APTES, would increase the formed Si-O-Al bonds on the expense of stronger epoxy-metal bonds resulting in a less adherent coat [2]. Also, increasing the ratio of APTES would force the formation of a thin layer of silica (Figure 2) with an unaviodable micro-cracks that further decrease the adhesive power of the hybrid coat [2].

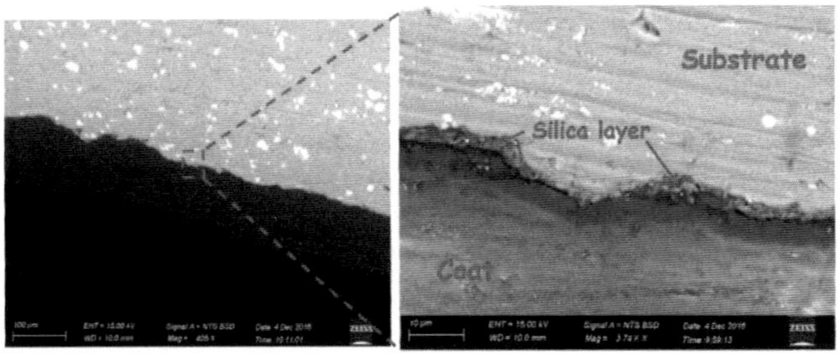

Figure 2. SEM images for a cross-section of the A5 sample showing silica agglomeration over the interface.

3.2.2 Hardness test

Applying Vickers hardness test on the HV10 mode to the different hybrids had come with the results listed in Table 1. The results showed an increase in the hardness with the increase in the silica content (adding APTES) reaching a max. at A2. Further addition would form silica particles in clusters. Silica agglomerations would show micro-cracks within the coat affecting largely the hardness [2]. Figure 3 shows SEM images for A4 and A5 revealing the micro-cracks found in silica agglomerations.

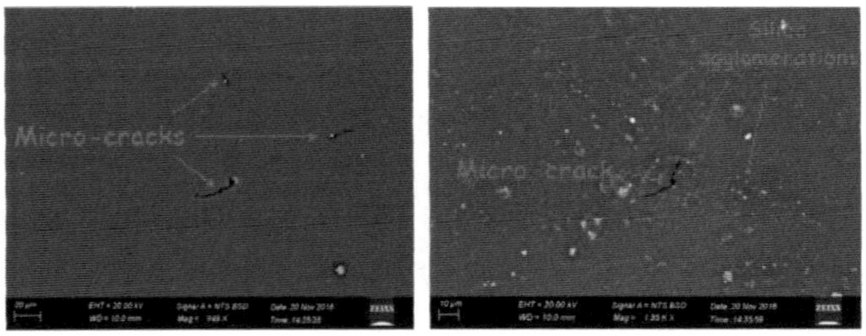

Figure 3. SEM images for A4 and A5 samples' surface showing fine micro-cracks.

3.3 Contact angle test

Applying a 1 μm distilled water drop, contact angle investigation has come out with the following results showed in Table 1. The results showed that the coat formed is generally hydrophilic and that the wettability of the surface coat varies along the different compositions. It was mentioned that the addition of silica particles would affect the cross-linking density of the hybrid network causing an increase in the free volume as shown in Figure 4. Using silane coupling agents, like GLYMO, would create covalent bonds between networks that decrease the free volume as much as possible [9]. Introducing APTES, would increase the bonding between both networks resulting in a more compact hybrid [10]. The results for (A0, A1, and A2) showed that the more the APTES is added, the more compact the network is, which results in an increase in the contact angle. In other words, the wettability decreases by the addition of the APTES reaching a minimum at the sample A2 then increases again.

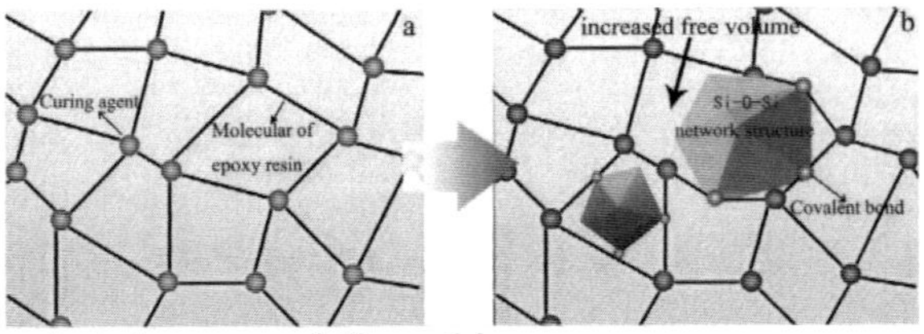

Figure 4. The increase in the free volume due to the presence of silica particles.

Further addition of APTES over the A2 sample, showed a gradual increase in the wettability (decrease in the contact angle) for higher silica content samples (A3, A4, and A5). This could be attributed to the increase of the stresses during the formation and curing of the hybrid network as a result of a high silica percentage [11]. The existence of such stresses would create micro-cracks within the coat that ease the diffusion of water through the coat making the coat more and more hydrophilic.

3.4 Thermal Gravimetric analysis (TGA)

TGA is used to evaluate the thermal stability and to determine the silica content involved for each hybrid composition. Combustion of a small amount of each prepared coat in a dynamic airflow from ambient temperature to 800°C would give the results shown in Table 1.The silica content for all samples is represented by the remaining weight with respect to that originally introduced (excluding the ash content from the neat epoxy). Figure 5 shows the TGA thermographs of the neat epoxy along with its silica-bonded hybrids. It is clear from the graphs that the neat epoxy sample has shown nearly a complete decomposition while introducing GLYMO had increased the silica content up to 12%. Using different compositions of APTES increases the inorganic content (represented in silica) of the coat. A completely APTES cured sample had shown the max. silica content of 27.6%.

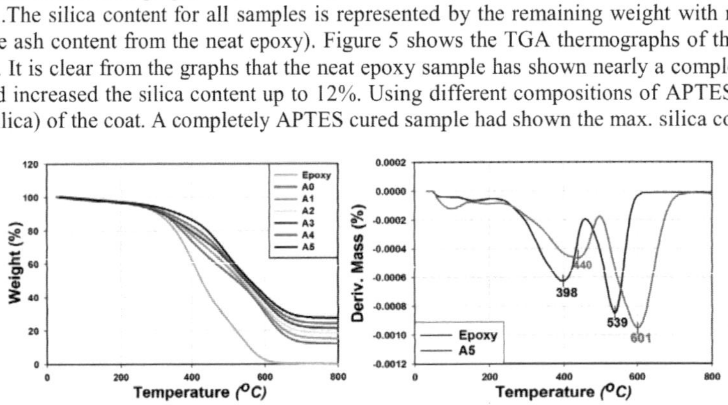

Figure 5. TG/DTG thermographs of the neat epoxy and the hybrid coats.

Reviewing the DTGA thermographs (Figure 5) shows that the thermo-oxidative degradation of epoxy and its hybrids is done over a three-stage process [12]. The curves starts with a small inclination at the range of (120-180oC) that corresponds to the first stage. In this stage, the loss of water adsorbed from the surrounding and that condensed from the residual hydroxyl groups of the silanol groups (Si-OH) are involved [2, 12, 13]. A second stage is associated to the breakage and the decomposition of the crosslinked organic polymer network in epoxy [2, 12]. Then, a third oxidative-degradation stage stems from the structural phase-separation conversion or an oxidation of the aromatic rings of epoxy [14,15].

A comparison of the DTGA of the DETA cured Epoxy and the A5 sample (the APTES cured (Epoxy/GLYMO) hybrid with the max. silica content) shown in Figure 5. It is obvious that a similar decomposition patterns are obtained that confirms the contribution of the organic components only to the weight loss in the hybrid materials. However, all degradation steps were found to occur at slower rates and higher temperatures for the silica-contained hybrids. The increased onset decomposition temperature (Td) from 398oC to 440oC is attributed to the insulating barriers created by the heat-resistant nano-silica particles on the matrix that protects the organic matrix from the thermo-oxidative degradation [12]. While the increase in the other decomposition temperature from 539oC to 601oC is attributed to the covalent bonds formed between the amino groups linked to the silica-networks and the epoxy resin that can postpone the breakdown of the aromatic rings [15]. Therefore, the introduction of silica to the thermoset epoxy coat had offered a stability to the all hybrids rather than the neat epoxy [1, 12, 16].

4 Conclusions

Silica-based hybrid coatings prepared by the reaction of DGEBA epoxy resin and GLYMO in the presence of APTES and DETA were prepared and characterized. Confidently, FTIR spectra confirmed the formation of silica domains within the epoxy matrix. It was observed that the adhesion strength has obviously risen with increasing the APTES content to that of the A2 sample

(APTES:DETA ratio of 1:1.5 by wt.) due to the increase in bonding with the substrate. Also, the same sample has recorded the highest hardness of 221.0 HV compared to other compositions as it contains the max. percentage of silica with no cracks within the coat. On the other side, increasing the silica content was observed to increase the thermal stability of hybrid coat, in which sample A5 (with silica content up to 27.59%) has shown higher thermal stability rather than the other hybrids and the neat epoxy. Sample A2 has shown the best-compromised performance among the prepared hybrids.

References

1. D. Levy and M. Zayat, eds. The Sol-Gel Handbook: Synthesis, Characterization and Applications, 3-Volume Set, (John Wiley & Sons, 2015)
2. E. Bakhshandeh, A.Jannesari, Z. Ranjbar, S. Sobhani, M. R. Saeb, Prog. Org. Coat. **77**, 1169-1183 (2014)
3. S. R. Davis, A. R. Brough, A. Atkinson, J. Non-Cryst. Solids **315**, 197-205 (2003)
4. J. Coates, EAC (2006)
5. V. Tagliazucca, *Doctoral dissertation*, University of Trento (2010)
6. I. M. Šapić, L. Bistričić, V. Volovšek, V. Dananić, Macromol. Symp. **339**, 122-129 (2014)
7. W. Zhang, A. A. Dehghani-Sanij, R. S. Blackburn, Prog. Nat. Sci. **18**, 801-805 (2008)
8. R. M. Almeida, C. G. Pantano, J. Appl. Phys. **68**, 4225-4232 (1990)
9. C. Peng, J. Li, Z. Wu, W. Peng, D. Zhou, RSC Adv. **6**, 38300-38309 (2016)
10. S. L. Case, *Doctoral dissertation*, Virginia Tech. (2003)
11. X-R. Deng, Q-H. Zhang, X-Y. Lei, W. Yang, H-H. Hui, J. Wang, J. Shen, J. Sol-Gel Sci. Technol. **73**, 242-249 (2015)
12. Z. Ahmad, F. Al-Sagheer J. Sol-Gel Sci. Technol. **72**, 334-343 (2014)
13. J. Z. Zheng, X. P. Zhou, X. L. Xie, Y. W. Mai, Nanoscale **2**, 2269-2274 (2010)
14. Y. Liu, Z. Du, C. Zhang, H. Li, Int. J. Polym. Anal. Charact. **11**, 299-315 (2006)
15. Z. Xi, H. Qingliang, G. Hongbo, J. Mat. Chem. **1**, 2886 (2013)
16. K. Y. Huang, C. J. Weng, S. Y. Lin, Y. H. Yu, J. M. Yeh, J. Appl. Polym. Sci. **112**, 1933-1942 (2009)

Study on Fuzzy Comprehensive Evaluation Model for the Safety of Mine Belt Conveyor

Xiaoyan Gong[1], Minxian Zhao[1a], Yue Wu[1] and Shaoni Qin[1]
[1]Xi'an University of Science and Technology, Xi'an, Shaanxi, 710054, China

Abstract. To improve the situation of the frequent failures of mine belt conveyor during operation, a model was used to evaluate the safety of mine belt conveyor. Based on the foundation of collecting and analyzing a large quantity of fault information of belt conveyor in the nationwide coal mine, the fault tree model of belt conveyor has been built, then the safety evaluation index system was established by analyzing and removing some secondary indicators. Furthermore, the weighted value of safety evaluation indexs was determined by analytic hierarchy process(AHP), and the single factor fuzzy evaluation matrix was constructed by experts grading method. Additionally, the model was applied in evaluating the security of belt conveyor in Nanliang coal mine. The results shows the security level is recognized to the "general", which means that this model can be adopted widely in evaluating the safety of mine belt conveyor.

1 Introduction

As a widely used and efficient transportation equipment, the stable operation of belt conveyor is very important to ensure the production safety of coal mine[1,2]. At present, most coal mine takes regular maintenance on the belt conveyor, but this method is unable to eliminate the hidden danger timely, which leads to frequent failures of conveyor and causes heavy casualties and damage to severe economic property[3]. There are increasing researchers initiate the study of monitoring system to improve the safety and reliability of conveyor, such as Li H J[4] who designs a slipping-belt protecting and monitoring system based on CAN and WSNs to real-time monitor the operating parameters to reduce the occurrence of slipping faults; Wang C[5] who devises a belt conveyor monitoring system based on PLC to master the running state of conveyor in real time, which ensures the safe and reliable operation of conveyor. However, the field investigation indicates that some hidden dangers still exist. To avoid the happening of conveyor accidents, a reasonable and effective safety evaluation method was created. For the types of conveyor faults and the uncertainty of relationship among them, AHP and expert scoring method were used to establish the fuzzy evaluation model of belt conveyor, which can quantify the evaluation indicators, and improve the accuracy of the evaluation results. Moreover, this model was be used in Nanliang coal mine, with the safety analysising of belt conveyor, we can find out the defections and come up with the corresponding countermeasures to improve the safe operation of conveyor.

[a] Corresponding author : zhaomx92@163.com

2 The establishment of the fault tree model

Based on the understanding of working principles and structure characteristics of belt conveyor, this paper extracts the main types of safety hazard of belt conveyor and confirms the logical relationship among faults, then two-level fault tree model of belt conveyor is established, which lays the foundation of the safety evaluation of belt conveyor. The fault tree model is shown in figure1.

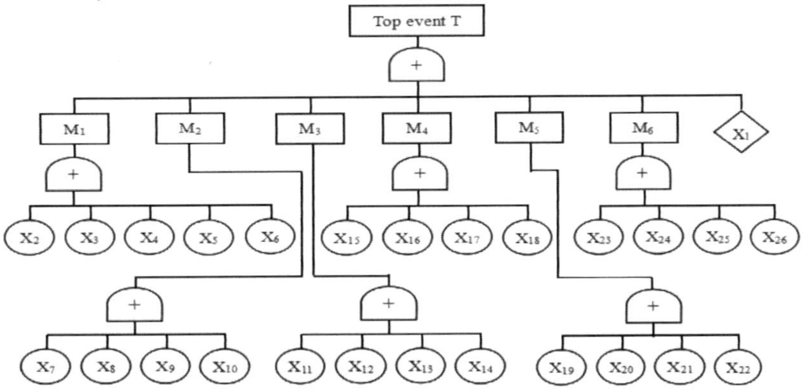

T: safety hazard of belt conveyor M_1: slipping fault M_2: tearing fault
M_3: smoking fault M_4: material scattering failure M_5: deviation fault
M_6: motor fault X_1: the lack of tension belt X_2: driving drum worn seriously
X_3: belt appeared thin and narrow X_4: roller damaged X_5: the belt is overload
X_6: belt damaged X_7: foreign bodies stuck in belt X_8: joint damaged and folding
X_9: the belt was broken for overload X_{10}: drum smoke for wear seriously X_{11}: belt smoke for wear
X_{12}: the temperature of roller is rise X_{13}: electrical short circuit lead to fire X_{14}: the scattered of coal
X_{15}: floating coal is too much X_{16}: guide groove damaged X_{17}: the belt is suspended
X_{18}: the belt center line is deviate X_{19}: aging belt appeared S X_{20}: the nose is uneven force
X_{21}: loading position is wrong X_{22}: shell lost heat is not enough X_{23}: motor fan damaged
X_{24}: fastening bolts are loose X_{25}: long time overload operation, motor heating lead to short circuit

Figure1. Fault tree model of belt conveyor

3 The establishment of fuzzy comprehensive evaluation model

3.1 The determination of the safety evaluation index system

Based on the fault tree model of belt conveyor, some secondary indicators which combined with survey data of conveyor was deleted to determine the safety evaluation index of belt conveyor. Then, the factor set of evaluation target is U={U1, U2, U3, U4, U5, U6}; sub factor set Ui={Ui1, Ui2, ...Uin} (i=1,2, 6); The safety evaluation index system of belt conveyor is shown in figure2.

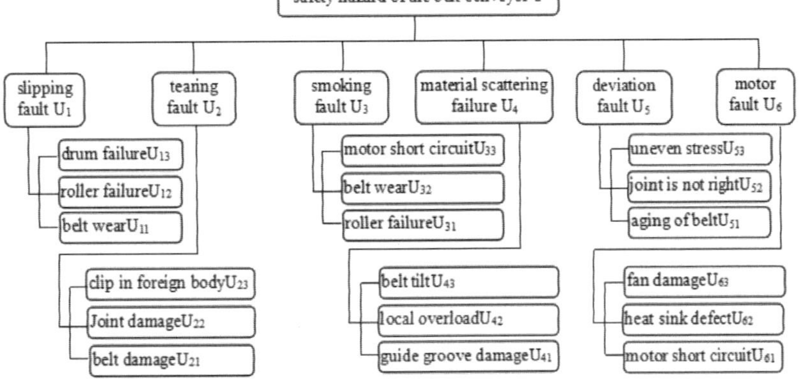

Figure2. safety evaluation index system of belt conveyor

3.2 The construction of the evaluation set

Through the statistical analysis of the number of failures and the causes of conveyor in the field investigation, the belt conveyor experts evaluate the safety level of belt conveyor according to the possibilities and severities of the accident occurrence. In this paper, the safety status of belt conveyor is divided into 5 grades, which is written as V={V1, V2, V3, V4, V5}={safe, relatively safe, general, relatively dangerous, dangerous}={>0.95, 0.95-0.90 0.90-0.85, 0.85-0.80, <0.80}.

3.3 The determination of the weight of factors based on AHP

3.3.1 Constructing pairwise comparison matrices

According to the Saaty 1~9[6]scale method(as shown in table1), the pairwise comparison matrix was constructed by comparing every two fuzzy factors of evaluation index system, through normalization, we got the weight vector Zi=(Zi1, Zi2,⋯,Zin), and $\sum_{j=1}^{n} z_{ij} = 1$

Table1. The 1~9 scale table

scale	definition	meaning
1	equally important	same contribution to target
3	slightly important	slightly different
5	suitable importance	different
7	really important	obvious difference
9	absolutely important	obvious importance
2,4,6,8	intermediate to two degrees	need to compromise
The reciprocal	If the element i compares element j that obtained the resulting bij, then the resulting of the elements j compares element i that is 1/bij	

3.3.2 Calculating the weight vector and checking the consistency

The maximum eigenvalue and the corresponding eigenvector were obtained according to the pairwise comparison matrix, then calculated the deviation consistency index CI ,which compared with average random consistency index RI[7], If the consistency check passed, the weight vector of evaluation factors set can be obtained by normalizing the eigenvector of evaluation matrix. Otherwise, we need to reconstruct the pairwise comparison matrix.

1 Calculate the maximum eigenvalue of the evaluation matrix

$$AW = \lambda_{\max} W \tag{1}$$

In the formula (1), $\lambda_{\max}$ is the largest eigenvalue of a matrix A, W is the eigenvector corresponding to the largest eigenvalue of matrix A.

2 Consistency check

The formula for calculating deviation consistency index:

$$CI = (\lambda_{\max} - n)/(n-1) \tag{2}$$

Average random consistency index: RI can be found from table 2.

The formula for calculating consistency:

$$CR = CI / RI \tag{3}$$

If CR<0.1,which indicates that the evaluation matrix passed the consistency test; otherwise, the contrast matrix ought to be reconstructed.

Table2. The standard value of average random consistency index RI

Matrix order	1	2	3	4	5	6	7	8	9
RI standard value	0	0	0.52	0.89	1.12	1.26	1.36	1.41	1.46

3.4 The establishment of a single factor matrix

Combined with their own experience and the influence degree of each index of belt conveyor, the experts and specified technicians were invited to vote according to the scoring criteria, then the single factor matrix was founded based on the principles of maximum subordinate.

$$R_i = \begin{bmatrix} r_{11} & r_{12} & \cdots & r_{1m} \\ r_{21} & r_{22} & \cdots & r_{2m} \\ \vdots & \vdots & \cdots & \vdots \\ r_{n1} & r_{n2} & \cdots & r_{nm} \end{bmatrix}$$

The formula of first level fuzzy comprehensive evaluation:

$$V_i = Z_i \bullet R_i = (b_{i1},\ b_{i2},\cdots,\ n_{im}) \tag{4}$$

In the Chad operator algorithm, the" $\wedge$ "symbol refers to take small operations, the" $\vee$ "symbol refers to take big operations, and for example:

$$b_j = \mathop{\vee}_{i=1}^{n} (z_i \wedge r_{ij}) \tag{5}$$

The formula of second level fuzzy comprehensive evaluation:

$$V = Z \bullet R = Z \bullet (V_1 \quad V_2 \quad \ldots \quad V_n)^T \tag{6}$$

4 The safety evaluation of belt conveyor in Nanliang coal mine

4.1 The calculation of weight vector

Based on the experience of the coal mine belt conveyor and some other criteria, relevant experts and field technicians construct pairwise comparison matrix by using the 1~9 scale method to calculate the weight vectors of the belt conveyor. Take the first weight vector as an example, the calculation is shown in table 3.

Table3. The calculation of the first weight vector.

U	U_1	U_2	U_3	U_4	U_5	U_6	K_i	weight vector Z_i	Consistency check
U_1	1	3	4	5	1/3	6	19.33	0.302	
U_2	1/3	1	3	2	1/5	2	8.53	0.133	
U_3	1/4	1/3	1	2	1/5	2	5.78	0.090	$\lambda_{max} = 6.41$
U_4	1/5	1/2	1/2	1	1/3	2	4.53	0.071	$CR = CI/RI =$
U_5	3	5	5	3	1	6	23.00	0.360	[(6.41-6)/5]/1.26=0.07
U_6	1/6	1/2	1/2	1/2	1/6	1	2.83	0.044	qualified
					$Z_i = k_i/\sum k_i$,		$\sum Z_i = 1$		

From table3, we got $CR = 0.07 < 0.1$, which means the first weight vector passes the consistency test, and the first weight vector can be written as Z=(0.302 0.133 0.090 0.071 0.360 0.044).

As well, the establishment of pairwise comparison matrix of the second level evaluation and the judgment matrices are shown in table 4.

Table4. Matrices of the the second level

U_1	U_{11}	U_{12}	U_{13}	U_2	U_{21}	U_{22}	U_{23}	U_3	U_{31}	U_{32}	U_{33}
U_{11}	1	2	3	U_{21}	1	3	5	U_{31}	1	1/3	2
U_{12}	1/2	1	3	U_{22}	1/3	1	3	U_{32}	3	1	3
U_{13}	1/3	1/3	1	U_{23}	1/5	1/3	1	U_{33}	1/2	1/3	1
U_4	U_{41}	U_{42}	U_{43}	U_5	U_{51}	U_{52}	U_{53}	U_6	U_{61}	U_{62}	U_{63}
U_{41}	1	2	3	U_{51}	1	1/3	2	U_{61}	1	3	5
U_{42}	1/2	1	3	U_{52}	3	1	3	U_{62}	1/3	1	3
U_{43}	1/3	1/3	1	U_{53}	1/2	1/3	1	U_{63}	1/5	1/3	1

Through calculation, the consistency of the two level weight vector is qualified, and the judgment matrix can be used. According to table4, the weight vectors of the two level are Z1=(0.575 0.311 0.114), Z2=(0.603 0.290 0.107), Z3=(0.274 0.575 0.151), Z4=(0.575 0.311 0.114), Z5=(0.274 0.575 0.151), Z6=(0.603 0.290 0.107).

4.2 Determination of single factor evaluation matrix

Based on the operation of the conveyor and experience, the experts and the field technicians' voting are accomplished by means of scoring, then the single factor matrix is established, and the fuzzy evaluation matrixes are obtained after normalization processing.

$$R_1=\begin{bmatrix}0.1&0.2&0.4&0.2&0.1\\0.2&0.3&0.2&0.2&0.1\\0.5&0.1&0.2&0.1&0.1\end{bmatrix}\quad R_2=\begin{bmatrix}0.1&0.2&0.4&0.2&0.1\\0.1&0.2&0.3&0.3&0.1\\0.4&0.2&0.2&0.1&0.1\end{bmatrix}\quad R_3=\begin{bmatrix}0.4&0.2&0.2&0.1&0.1\\0.1&0.2&0.3&0.3&0.1\\0.4&0.3&0.2&0.1&0\end{bmatrix}$$

$$R_4=\begin{bmatrix}0.1&0.2&0.3&0.3&0.1\\0.4&0.2&0.2&0.1&0.1\\0.2&0.3&0.2&0.2&0.1\end{bmatrix}\quad R_5=\begin{bmatrix}0.1&0.2&0.3&0.3&0.1\\0.2&0.2&0.4&0.1&0.1\\0.3&0.3&0.2&0.1&0.1\end{bmatrix}\quad R_6=\begin{bmatrix}0.1&0.2&0.3&0.3&0.1\\0.4&0.2&0.2&0.1&0.1\\0.3&0.3&0.2&0.2&0\end{bmatrix}$$

4.3 Calculation and evaluation the safety of conveyor

Based on the operation of the conveyor and experience, the experts and the field technicians' voting are accomplished by means of scoring, then the single factor matrix is established, and the fuzzy evaluation matrixes are obtained after normalization processing.

4.3.1 The first comprehensive evaluation

Under the condition that the single factor fuzzy evaluation matrix and the weight vector value are determined, we can calculate according to formula3.

$$V_1=Z_1\bullet R_1=(0.575\quad 0.311\quad 0.114)\times\begin{bmatrix}0.1&0.2&0.4&0.2&0.1\\0.2&0.3&0.2&0.2&0.1\\0.5&0.1&0.2&0.1&0.1\end{bmatrix}.$$

According to the rule of Chad operator, we can got V_1=(0.200 0.300 0.400 0.200 0.100), similarly, V_2=(0.107 0.200 0.400 0.290 0.100), V_3=(0.274 0.200 0.300 0.300 0.100), V_4=(0.311 0.200 0.300 0.300 0.100), V_5=(0.200 0.200 0.400 0.274 0.100), V_6=(0.290 0.200 0.300 0.300 0.100).

4.3.2 The second comprehensive evaluation

The evaluation vector obtained from the first level evaluation is combined into the evaluation matrix of the second level evaluation R. The second comprehensive evaluation is following:

$$V=Z\bullet R=(0.302\quad 0.133\quad 0.090\quad 0.011\quad 0.360\quad 0.044)\times\begin{bmatrix}0.200&0.300&0.400&0.200&0.100\\0.107&0.200&0.400&0.290&0.100\\0.274&0.200&0.300&0.300&0.100\\0.311&0.200&0.300&0.300&0.100\\0.200&0.200&0.400&0.274&0.100\\0.290&0.200&0.300&0.300&0.100\end{bmatrix}$$

$$=(0.200\quad 0.300\quad 0.360\quad 0.274\quad 0.100)$$

After normalization, we got the V=(0.162 0.243 0.292 0.222 0.081), according to the principle of maximum membership degree, the corresponding level of the largest number in the fuzzy evaluation set V is the level of the evaluation. The results show that the safety level of the belt conveyor is grade 3,which means the safety of belt conveyor is "general", and its safety needs to be improved.

From the table 3 we can see, the weight value of slip fault and deviation fault up to 0.302 and 0.360 ,which indicates the failure is relatively large for slip fault and deviation. The slip fault belt is prone to wear, it is mainly because that the conveyor belt has to carry large amounts of coal and the edge of coal is not even, the belt will often be worn. In order to prevent slipping failures, relevant measures should be taken to resolve the problem. Such as checking the conveyor belt is clean or not, checking the belt condition of deformation and aging problems regularly, cleaning or replacing the conveyor belt timely, reducing the quantity of conveyor transportation and so on. The deviation fault is likely to be caused by the wrong joint of belt, which can be found by checking whether the belt joint appears skin cracks and the edge burr, therefore we can readjust or remake joint according to requirements, and clean air conveying section which ensure the clean of roller surface to reduce the occurrence of accidents.

5 Conclusion

Aiming at the situation that frequent failures appear on the belt conveyor, a fuzzy comprehensive evaluation model of belt conveyor safety is established. In this paper, we use AHP to determine the weight of each index value and the expert scoring method to build up single factor fuzzy evaluation matrix, which combines the qualitative analysis and quantitative evaluation to reflect the safety degree of belt conveyor objectively. A case study of Nanliang coal mine was conducted based on this model, the results show that its security level is "general", and the slipping fault and deviation fault are the main factors that influence the safety of mine belt conveyor, we can put forward the corresponding countermeasures to improve the safety operation of conveyor. In addition, the model can be combined with the actual operation of the conveyor to build the corresponding evaluation matrix for specific coal mines. Thus this model has practical application values in improving the safety of mine belt conveyor.

References

1. W J.Li. Fault tree analysis of mine belt conveyor. Modern Mining, **7**, 179-180 (2015)
2. B.Luo. Study on fault diagnosis intelligent decision support system of mine belt conveyor based on information fusion, Xi'an University of Science and Technology, (2015)
3. Y Y.Cui. Fault diagnosis and countermeasure of telescopic belt conveyor. Hebei University of Engineering, (2013)
4. H J Li, S G Miao. Research of strap skip protecting and monitoring system in coal mine based on CAN and WSNs. Transducer and Microsystem Technologies, **30**(12) ,40-42 (2011)
5. C.Wang. Research of long-distance monitor and control system based on PLC for belt conveyor in coal mine. Liaoning Technical University, (2011)
6. Y.Chang, X D.Liu. On application of AHP and fuzzy evaluation in enterprise's technologic innovation capability. Science & Technology Progress and Policy, **19** (9), 125-127. (2002)
7. W F.Cai, H B.Wang. The risk assessment of mine emergency management based on AHP-FCE comprehensive evaluation. Science and Technology Management Research, **16**, 30-33 (2013)

DOI: 10.1051/matecconf/201713602002

Numerical Investigation on the Flow Field Produced by Oscillating Foils

Han-Chieh Chiu[1], Ren-Horn Shieh[1], Yu-Jen Chiu[1], Jer-Huan Jang[2,3,a], and Cheng-Ru Yu[2]

[1]*Department of Mechanical Engineering, Taipei City University of Science and Technology, Taipei 112, Taiwan*
[2]*Department of Mechanical Engineering, Ming Chi University of Technology, New Taipei City 243, Taiwan*
[3]*Department of Mechanical Engineering, Chang Gung University, Taoyuan 333, Taiwan*

Abstract. In present manuscript, the flow field produced by oscillating foils in a duct were under investigation numerically. The flow field was considered as an incompressible laminar transient flow. In this paper, the oscillating frequencies were 2, 4 and 6 Hz; the span angles of the foil were 30°, 60°, 90°, 120°, and 150°, respectively. The flow field produced by a particular oscillating pattern (fast forward slow backward, FFSB) has been made in comparison with that by the sinusoidal oscillating pattern. The flow field produced by dual oscillating foils was also analysed. Two foils oscillated in the same direction (in-phase) and in the opposite direction (counter-phase) were studied as well. Commercial software ANSYS-FLUENT was employed for the 2D numerical simulation. A moving grid technique was utilized in the analysis. Results present the flow average velocity increases as the oscillating frequency increases. It is also shown that the flow average velocity with span angle of 120° is the largest among these cases. In addition, the flow average velocity with dual foils is better than that with a single foil.

1 Introduction

The progress of semiconductor industry has reduced the scale of electronic devices and increases the heat flux of heat sources. The heat dissipation rate from electronic devices requires new designs of miniaturized cooling components. Due to limitations of space in electronic devices, forced convective cooling with fans was not available. However, electronic cooling with combined conduction and free convection was hardly sufficient for electronic devices. One of the potential issues is the convective heat transfer with miniature oscillating foils. Therefore, research on the oscillation of small plate become one of the important issues for compact electronic cooling.

In cases where rotary fans are ineffective for dissipating heat, the piezoelectric fan provides a potential and reliable alternative approach to heat transfer. Piezoelectric fans can be fabricated in small sizes to fit applications that require miniature cooling modules. Toda [1, 2] first proposed the concept of piezoelectric fans for a cooling system with small size and low power consumption. He employed piezoelectric material as driven element with driving voltage of 110V and the frequency of 13Hz. As the plate was oscillating in the resonance frequency, its amplitude would increase and the surrounding air was driven greatly. The oscillating plate could be miniaturized and therefore considered as the cooling of small electronic devices with considerable potential.

Yoo et al [3] studied the resonant frequencies for various foil designs. The resonant frequency was found to be inversely proportional to square of foil length. The maximum flow velocity close to the tip of foil can reach 3.0 m/sec at appropriate resonant frequency. Wait et al [4] measured the amplitudes of piezoelectric fan for higher resonance modes. The energy consumption and flow field were assessed for various foil lengths and resonant frequencies. Acikailn et al. [5] observed the flow field around fan tip and investigated the heat transfer coefficient on heat source. Liu et al [6] tested the thermal performance of fans with various fan geometries and horizontal/vertical arrangements. Kimber and Garimella [7] measured various piezoelectric fans mounted perpendicular to a constant heat flux surface and studied the thermal performance across specific range of fan dimensions, vibration frequency, and amplitude. The experimental results show that vibration frequency has more significant effect on the heat transfer rate than vibration amplitude. Li et al [8] studied the effects of configuration, foil location, and heat sink design.

Fan array was a further development on this issue. Multiple fans were utilized in driving air flow and cooling system. Kimber and Garimella [9] built dual fan array to study the cooling performance on vertical surface. Ma et al. [10] designed multiple vibrating-fan cooling system actuated by combining piezoelectric force and magnetic force.

From the literatures cited above, it is found that most of the studies are investigated through experiment. Numerical analysis on the flow field induced by oscillating foil has been rarely studied. The main purpose of this paper is to investigate the air flow induced by single and dual oscillating foils. The effects of oscillating frequency, and oscillating pattern, span angle of oscillation have been studied to estimate the average velocity of the air flow.

[a] Corresponding author: jhjang@mail.mcut.edu.tw

 DOI: 10.1051/matecconf/201713602002

2 Numerical analysis

2.1 Physical Model

The physical model of the present study is shown in Figure. 1. The channel length is 400 mm with width of 90 mm and depth of 5 mm. An oscillating foil is located 80 mm away from the inlet and at the centre of the channel, as shown in Figure. 1(a). The foil is rectangular and 40 mm in length with thickness of 1.5 mm and height of 4.5 mm. The span angle of foil is 20 with oscillation symmetric to the centreline. The span angles of the foil are 30°, 60°, 90°, 120°, and 150°, respectively. In present study, dual foil oscillation is also under investigation. The second foil is located downstream to the first one. The distance between two foils is varied to study its influence on the flow. The distance between two foils are 44 mm, 50 mm, 60 mm, 70 mm, and 80 mm, respectively. The two foils are in the same dimension.

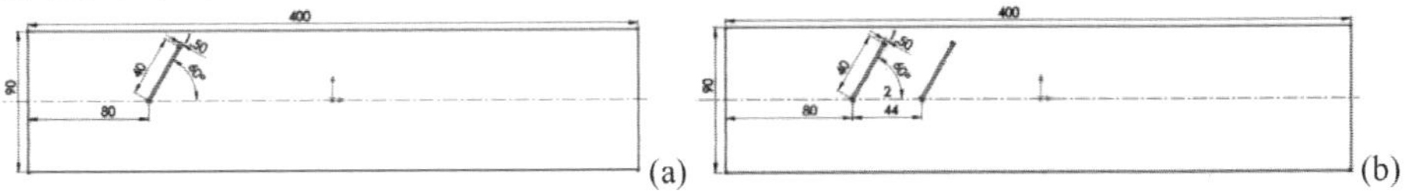

Figure 1. Two-dimensional physical model with (a) single foil; (b) dual foils.

2.2 Analysis

Because of the physical model, the depth of the channel is relatively small compared with its length and width, the air flow is considered as two-dimensional flow. Due to small amount of air drawn into the channel, the induced air flow is assumed to be incompressible laminar flow with constant properties in the present model. Actually, the foil would deform due to eccentric force by oscillation. However, the foil is considered rigid in order to simplify the problem.

The commercial software of ANSYS-Fluent was employed to analyse the physical 2D model. The dynamic mesh scheme was adopted in the area of foil sweeping, the meshes are defined as deformable meshes, such that they can be stretched or compressed while the foil is in motion. Therefore, the elements around oscillating foils would be stretched or compressed as shown in Figure. 2. As for the boundary conditions, oscillating foil and the wall were set as wall condition, the inlet was set as vent inlet, and the outlet was set as vent outlet. A transient flow with oscillating foils in motion for 10 s has been calculated. There are 20 steps for one complete cycle.

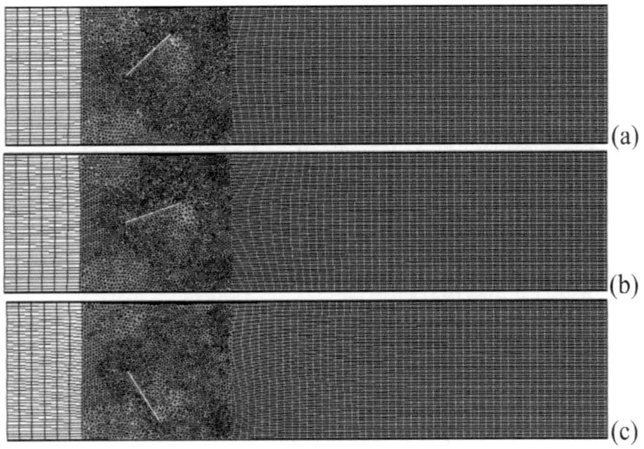

Figure 2. Sketch diagram of the dynamic meshes.

Most of the researches considered oscillating velocity symmetric to the central axis of the oscillation. However, the optimal oscillating velocity profile may not be in the sinusoidal form or a harmonic function. From the aspect of fluid dynamics, the rotating velocity should be higher as the foil is sweeping toward downstream. In other word, the rotation of foil from one end to the central line could be faster than that from the central line to the other end. In this paper, a fast forward slow backward (FFSB) oscillating pattern has been generated to investigate the effect of oscillating pattern on the induced air flow. In the FFSB pattern, as the foil rotates from one end to the central line, the angular velocity is higher; while the foil rotates from the central line to the other end, the angular velocity is lower. The oscillating pattern can be generated in the user defined function in ANSYS-Fluent. The oscillating patterns are presented in Figure. 3.

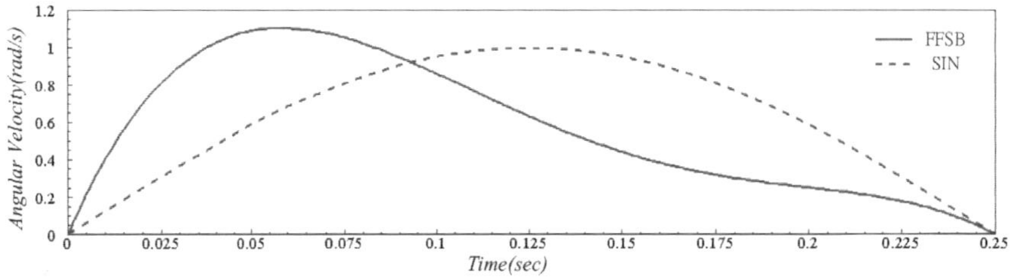

Figure 3. Oscillating pattern as a function of time for oscillating frequency of 2 Hz.

3 Results and Discussions

In present study, numerical simulation on the induced air flow by oscillating foils has been conducted. Effects of oscillating frequencies, span angles, oscillating patterns, and dual foils are considered. The oscillating frequencies are 2, 4, and 6 Hz; the span angles of the foil were 30°, 60°, 90°, 120°, and 150°, respectively. The flow field produced by dual oscillating foils was also analysed. Two foils oscillated in the same direction (in-phase) and in the opposite direction (counter-phase) were studied. The distance between two foils is varied to study its influence on the flow. The distance between two foils are 44 mm, 50 mm, 60 mm, 70 mm, and 80 mm, respectively.

3.1 streamline, velocity, and pressure distributions of flow field

Figure 4 shows the streamline diagram of the air flow with single oscillating foil at the frequency of 4 Hz and span angle of 120°. Figure 4(a) is the flow streamline as the foil is at the bottom; while Figure 4(c) is the flow streamline as the foil is at the centre. It is found that the air is drawn into the channel uniformly. In addition, vortices continuously occur towards channel outlet after the air flow passing through the oscillating foil. The corresponding velocity and pressure contours are presented in Figure. 5 and 6, respectively. It is seen that the high velocity occurs where stream lines are dense; while the pressure is low. It is also notice that vortices move downward regularly. However, some vortices are affected between the foil and the channel walls.

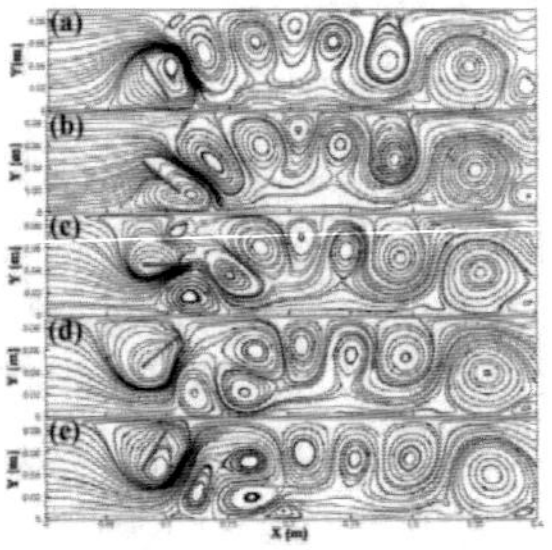

Figure4. Streamline diagram with single oscillating foil.

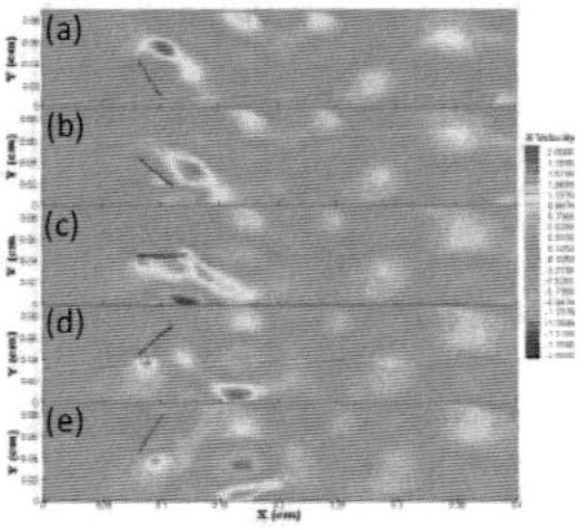

Figure 5. Velocity contour with single oscillating foil with oscillating frequency of 4Hz.

 DOI: 10.1051/matecconf/201713602002

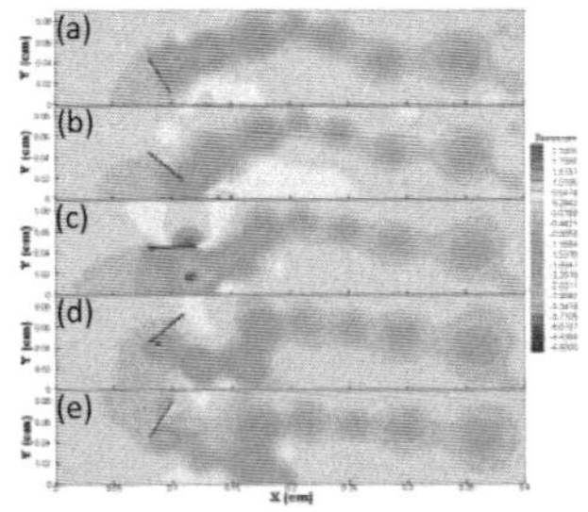

Figure6. Pressure contour with single oscillating foil with oscillating frequency of 4Hz.

3.2 Effect of span angle

The air flow induced by the oscillating foil is affected by the span angle of the foil. In this study, the effect of span angle on the averaged velocity in the channel has been investigated. The span angles are 30°, 60°, 90°, 120°, and 150°, respectively. Figure 7 presents the effect of oscillating span angle on the average velocity in the channel for three oscillating frequencies. It is found that the flow average velocity increases as oscillating frequency increases at each span angle. It is also discovered that the optimum flow average velocity occurs at the span angle of 120°. It is observed from the numerical flow animation that the flow velocity was hindered by the channel wall, resulted in a lower flow average velocity in the channel.

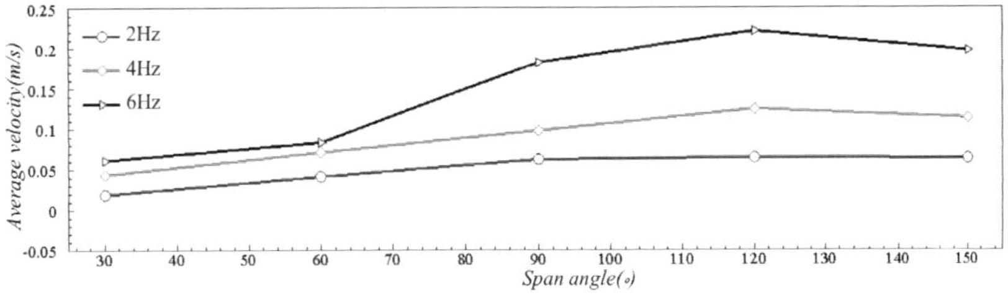

Figure 7. Effect of span angle on the flow average velocity in the channel for three oscillating frequencies.

3.3 Effect of oscillating pattern

Figure 8 depicts the effect of oscillating pattern on the flow average velocity in the channel for three oscillating frequencies. It is discovered that the flow average velocity with FFSB pattern is larger than that with sinusoidal pattern for three oscillating frequencies. This indicates that sinusoidal oscillating pattern is not an optimal oscillating pattern in producing air flow. This confirms the concept that a FFSB is a better oscillating pattern for producing air flow with oscillating foil.

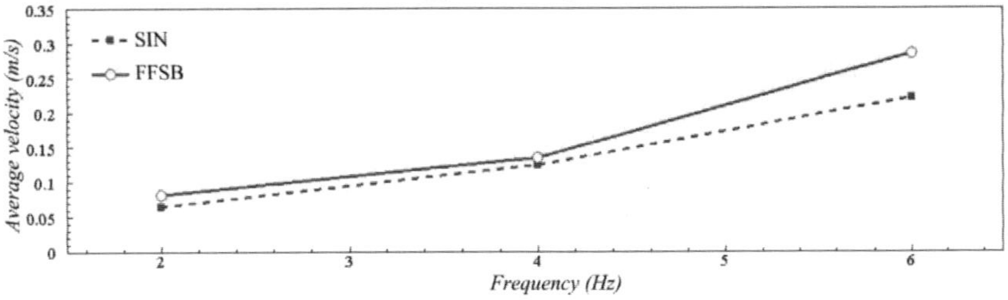

Figure 8. Effect of oscillating pattern on the flow average velocity in the channel for three oscillating frequencies.

3.4 Dual oscillating foils

In the present study, the air flow induced by dual foil oscillation was studied as well. In addition, two foils oscillated in the same direction (in-phase) and in the opposite direction (counter-phase) were studied. Figure 9 presents the effect of distance between two foils on the flow average velocity in the channel with three oscillating frequencies for both in-phase and counter-phase pattern. It is found that dual oscillating foils generate higher air flow than a single foil. In addition, it is obviously observed that the flow average velocity for counter-phase pattern is always larger than that for in-phase pattern. As for the effect of distance between two foils, it shows an optimum in the counter-phase pattern. However, the optimum condition does not appear in the in-phase pattern.

DOI: 10.1051/matecconf/201713602002

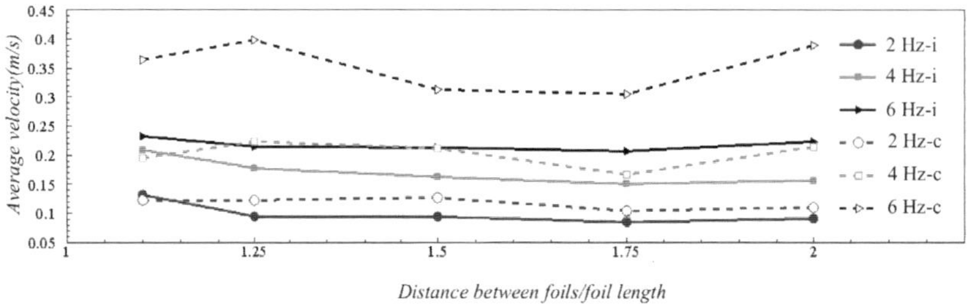

Figure 9. Effect of distance between two foils on the averaged velocity in the channel with three oscillating frequencies for both in-phase and counter-phase pattern.

4 Conclusions

In this paper, the air flow driven by oscillating foils in a channel has been numerically studied in detailed. Brief summaries are listed below:

1. Oscillating foils with higher frequencies generate higher flow average velocity than those with lower frequencies.
2. The effect of span angle on the flow average velocity in the channel has been investigated. It is discovered that the optimum flow average velocity occurs at the span angle of 120°.
3. It is discovered that the flow average velocity in the channel is larger than that FFSB pattern with FFSB pattern is larger than that with sinusoidal pattern for three oscillating frequencies.
4. In the oscillating range, it is observed that dual oscillating foils generate higher air flow than single foil. In addition, dual foil with counter-phase oscillating mode produces higher velocity air flow than that with in-phase oscillating mode. Besides, it shows an optimum in the counter-phase oscillating mode. However, the optimum condition does not appear in the in-phase oscillating mode.

Acknowledgement

The financial support of this research by the Ministry of Science and Technology under the Contract MOST 104-2221-E-149-002, MOST 105-2632-E-131-002, MOST 105-2632-E-131-003 is appreciated. Also, the support from Taipei City University of Science and Technology is acknowledged.

References

1. M. Toda, Ferroelectrics **22**, 911 (1979)
2. M. Toda, Ferroelectrics **32**, 127 (1981)
3. J.H. Yoo, J.I. Hong, W. Cao, Sensor Actuator A-Phys **79**, 8 (2000)
4. S.M. Wait, S. Basak, S.V. Garimella, A. Raman, IEEE Trans. Compon. Packag. Technol. **30**, 119 (2007)
5. T. Acikalin, S.M. Wait, S.V. Garimella, A. Raman, Heat Transfer Eng. **25**, 4 (2004)
6. S.F. Liu, R.T. Huang, W.J. Sheu, C.C. Wang, Int. J. Heat Mass Tran. **52**, 2565 (2009)
7. M.L. Kimber, S.V. Garimella, Int. J. Heat Mass Tran. **52**, pp. 4470 (2009)
8. H.Y. Li, S.M. Chao, J.W. Chen, J.T. Yang, Int. J. Heat Mass Tran. **57**, pp. 722 (2013)
9. M. Kimber, S.V. Garimella, ASME J. Heat Tran. **131**, 111401 (2009)
10. H.K. Ma, L.K. Tan, Y.T. Li, C.L. Liu, Int. Comm. Heat Tran. Vol. 55, pp. 77 (2014)

DOI: 10.1051/matecconf/201713602003

Analysis of the Working Cable System of Single-span Circulating Ropeway

Jian Qin[1], Liang Qiao[1], Jun Chen[1], Jiancheng Wan[1], Ming Jiang[1] and Chunhua Hu[1]
[1]*China Electric Power Research Institute, 100055, Beijing, China*

Abstract. Based on the analysis of the mutual interaction between bearing cable and traction cable of single-span circulating ropeway, the calculation method is proposed for working cables of ropeway with arbitrary loads. The 5 sets nonlinear equations are set up for the working cables according to the equal spans and altitude differences of the bearing cable and the traction cable, the constant of the load running space and the bearing cable length, and the tension balance of bearing cable and traction cable beside load. The resolving method is proposed based on the Newton iterative method. The correctness and feasibility of the method are proved by the results of engineering test. From the solution, the tension and coordinate of working cables can be obtained accurately, which is meaningful for the design and analysis of the freight ropeway.

1 Single-span freight ropeway

Aerial freight ropeway has strong adaptability for natural terrain and perfect climbing ability, so it can easy across mountains and rivers. It is widely used in forestry, construction, mine and other industries [1-3].

For the single-span ropeway, the accurate calculation of bearing cable tension is the key of the design and selection in ropeway system [4]. In the past, only the influence of weights of cable and loads are considered during the tension calculation of bearing cable [5-6], and the mutual interaction between bearing cable and traction cable is always ignored which will make the big error in tension calculation of the bearing cable. In this paper, a calculation method is proposed of mutual interaction between the bearing cable and the traction cable based on the elastic catenary theory.

2 Ropeway with arbitrary loads

The ropeway mainly includes working cables, brackets, trolleys and tractor. Working cables includes bearing cable and traction cable. The bearing cable is the important part of ropeway, which bears the loads of trolleys and cargos. The traction cable is a closed rope and driven by the tractor to move along the bearing cable in the ropeway [7].

(1) Parameters of ropeway

The single-span circulating ropeway is shown in Figure 1. The span between brackets is L_s, and altitude difference is h_s.

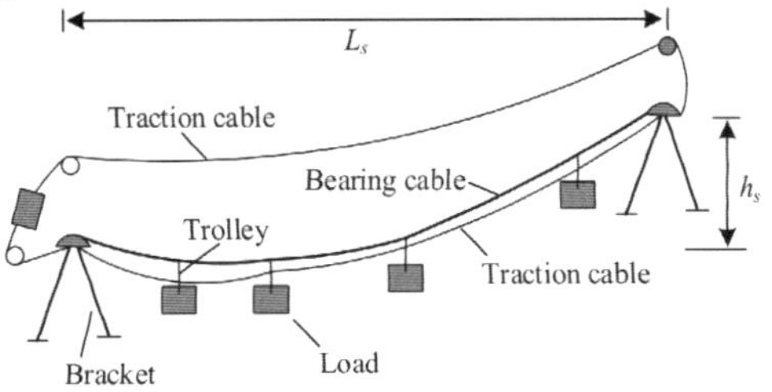

Figure 1. The single-span circulating ropeway

The length of bearing cable s^l_0 and length of traction cable s^t_0 can be measured after the erection of ropeway.

(2) Parameters of loads

There are m loads on the ropeway with the weights G_i (i=1,…, m) respectively. The lengths of traction cable between loads are s^G_i (i=1,…, m-1), which are kept constant during movement.

The length of traction cable from the tractor to the first load is defined as s^G_0 and the length between the last load and tractor is defined as s^G_m, shown as Figure 2.

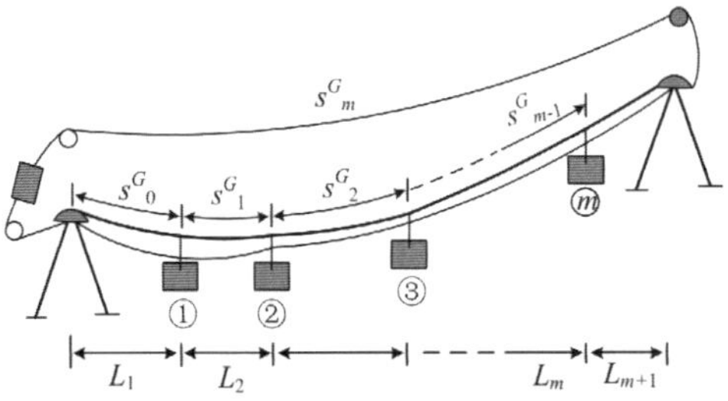

Figure 2. The single-span circulating ropeway

The working cables are divided by centralized loads as several catenary sections which only bear the self-weight of the cable. The beginning of every section is A, the ending is B. So there are m+1 bearing cable sections, and m+2 traction cable sections.

(3) Formulas of section

Assume the initial length of every section cable is s_i, horizontal tension is H_i, vertical tension of A is V_{Ai} (i=1,..., m+1), the vertical tension of B can be expressed as:

$$V_{Bi}=V_{Ai}+qs_i \quad (i=1,\ldots, m+1) \tag{1}$$

Here q is the unit length weight of cable.

The spans and altitude differences of every section are[7]:

$$\begin{aligned} L_i &= \frac{H_i}{q}\left[\frac{qs_i}{EA_0} + \operatorname{asinh}\frac{V_{Bi}}{H_i} - \operatorname{asinh}\frac{V_{Ai}}{H_i}\right] \\ h_i &= \frac{V_{Bi}^2 - V_{Ai}^2}{2qEA_0} + \frac{H_i}{q}\left(\sqrt{1+\frac{V_{Bi}^2}{H_i^2}} - \sqrt{1+\frac{V_{Ai}^2}{H_i^2}}\right) \end{aligned} \quad (i=1,\ldots,m+1) \tag{2}$$

Here EA_0 is the product of section area and elasticity modulus.

These variables above can represent bearing cable or traction with the superscripts l or t respectively.

3 Analysis of bearing cable and traction cable

The mutual interaction between bearing cable and traction cable of single-span ropeway is mainly reflected from the following aspects:

(1) The equal span and altitude difference of bearing cable and traction cable

As shown in Figure 2, the points A and B coincide in every section of bearing cable and traction cable that means the span and altitude difference of every section are equal. So the 2(m+1) equations below are obtained:

$$\begin{aligned} L_i^l - L_i^t &= 0 \\ h_i^l - h_i^t &= 0 \end{aligned} \quad (i=1,\ldots, m+1) \tag{3}$$

(2) Constant of the total span and altitude difference

No matter where the loads move in single-span ropeway, the sum of section spans is equal to whole horizontal span and the sum of section altitude difference is equal to whole altitude difference.

$$\begin{aligned} \sum_{j=1}^{m+1} L_j^l - L_s &= 0 \\ \sum_{j=1}^{m+1} h_j^l - h_s &= 0 \end{aligned} \tag{4}$$

For return traction cable, the equations are obtained as:

$$\begin{aligned} L_{m+2}^t - L_s &= 0 \\ h_{m+2}^t - h_s &= 0 \end{aligned} \tag{5}$$

(3) Tension balance of bearing cable and traction cable beside load

The bearing cable and traction cable are connected by the loads, shown as Figure 3(a). The number of cables beside load is i, i+1 respectively.

The wheel of trolley rolls on the bearing cable and makes cable bent down. Therefore, the tangential tensions of bearing cable beside load are equal, shown as Figure 3(b). So there are m equations of tension balance:

$$T_{Bi}^l - T_{Ai+1}^l = 0 \; (i=1,\ldots, m) \tag{6}$$

DOI: 10.1051/matecconf/201713602003

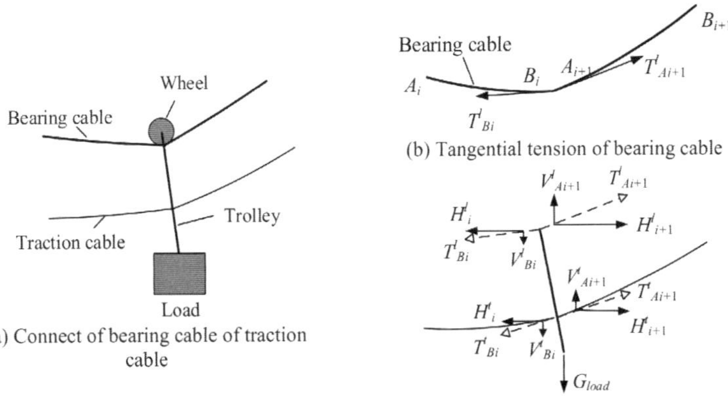

(a) Connect of bearing cable of traction cable

(b) Tangential tension of bearing cable

(c) Force analysis of trolley

Figure 3. Force analysis of bearing cable and traction cable beside load

The Force analysis of trolley is shown in Figure 3(c), the 2m tension balance equations in the horizontal and vertical directions can be obtained:

$$\begin{aligned} &H^l_{i+1} - H^l_i + H^t_{i+1} - H^t_i = 0 \\ &V^l_{Ai+1} - V^l_{Bi} + V^t_{Ai+1} - V^t_{Bi} - G_i = 0 \end{aligned} \quad (i= 1,\dots, m) \tag{7}$$

(4) Space constant of loads

The space between loads pulled by traction cable is unchanged during the movement. So we have m+1 equations.

$$\sum_{j=g_i+1}^{g_{i+1}} s^t_j - s^G{}_i = 0 \; (i=0,\dots,m-1) \tag{8}$$

and

$$\sum_{j=g_m+1}^{2+m} s^t_j - s^G{}_m = 0 \tag{9}$$

(5) Length constant of bearing cable

For single-span ropeway, the length of bearing cable has remained unchanged.

$$\sum_{j=1}^{m+1} s^l_j - s^l_0 = 0 \tag{10}$$

4 Solution of the nonlinear equations of working cable

Taking the section length s^l_i, s^t_i, horizontal tension H^l_i, H^t_i, vertical tension V^l_{Ai}, V^t_{Ai} (i=1, …, m+1) of the bearing cable and traction cable as the unknowns, the nonlinear equation set of single-span ropeway can be established based on interaction relationship of the working cable.

The number of unknown variables is 6m+9 and equal to the nonlinear equations number. Therefore, the nonlinear equations are closed, which can be expressed as:

$$F(X)=0 \tag{11}$$

In which $X=[s^l_1, \dots, s^l_{m+1}, H^l_1, \dots, H^l_{m+1}, V^l_1, \dots, V^l_{m+1}, s^t_1, \dots, s^t_{m+2}, H^t_1, \dots, H^t_{m+2}, V^t_1, \dots, V^t_{m+2}]^T$.

Newton iterative method can be used to solve the nonlinear equation set. Considered that Newton method is sensitive to the initial value, the parabolic analytical solutions is used as the initial iteration value.

(1)The length s^t_i (i=1,…,2+m) of every section divided by load point can be obtained according to the whole traction cable length and load distribution. Every section length of bearing cable is calculated in proportion.

$$s^{l0}_i = \frac{s^l_0}{s^t_0} s^t_i \quad (i=1, \dots, 1+m) \tag{12}$$

(2)According to bearing cable length s^l_0, the sag can be calculated basis of parabolic equation:

$$f = \sqrt{\frac{3(s^l_0 - L_s)}{8L_s}} \tag{13}$$

And then the horizontal tension of every section is obtained:

$$H^l = \frac{q^l L_s^2}{8f} \tag{14}$$

It is known from the balance conditions that the initial horizontal tension of every section is same.

$$H^{l0}_1 = H^{l0}_2 = \cdots = H^{l0}_{m+1} = H^l \tag{15}$$

 DOI: 10.1051/matecconf/201713602003

(3)The initial vertical tension of point A of first section can be obtained basis of sag and span.

$$V_1^{l0}=\frac{4f+h_s}{L_s} \tag{16}$$

Other sections' vertical tension can be calculated refer to equation below:

$$V_{i+1}^{l0}=V_i^{l0}+q^l s_i^{l0}+G_i \qquad (i=1,\ \ldots,\ m) \tag{17}$$

So the initial iteration value of bearing cable is:

$$X^{l0}=[\ s^{l0}{}_1,s^{l0}{}_2,\ldots,s^{l0}{}_{1+m},H^{l0}{}_1,H^{l0}{}_2,\ldots,H^{l0}{}_{1+m},V^{l0}{}_1,\ldots,V^{l0}{}_{1+m}]. \tag{18}$$

And the initial iteration value of traction cable can be obtained in the same way:

$$X^{t0}=[\ s^{t0}{}_1,s^{t0}{}_2,\ldots,s^{t0}{}_{1+m},H^{t0}{}_1,H^{t0}{}_2,\ldots,H^{t0}{}_{1+m},V^{t0}{}_1,\ldots,V^{t0}{}_{1+m}]. \tag{19}$$

The initial iteration value of the nonlinear equations is:

$$X^0=[\ X^{l0},X^{t0}]^T. \tag{20}$$

After X is calculated, the vertical tension V_x, tangential tension T_x and the horizontal and vertical coordinates of any point of the working rope at length s_x can be obtained respectively.

5 Example

In the single-span cableway experiment, L_s=218.18m, h_s=118.66m, $EA^l{}_0$=25MN, q^l=1.6kg/m, $EA^t{}_0$=9MN, q^t=0.8kg/m. The length of bearing cable is 249.12m.

The first load case is G_1=850kg, The 2nd case is G_1=850kg G_2=425kg, and $s^G{}_1$=21m.

In the case 1, the load moves from the 11m to 87m. The measured and calculated tangential tension of initial point with the load position is shown in Figure 4(a). In the case 2, the load moves from the 11m to 83m. The relation is shown in Figure 4(b).

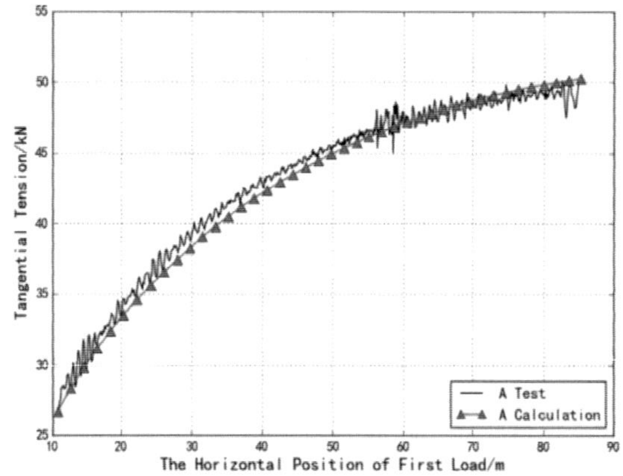

(a) Result of case 1

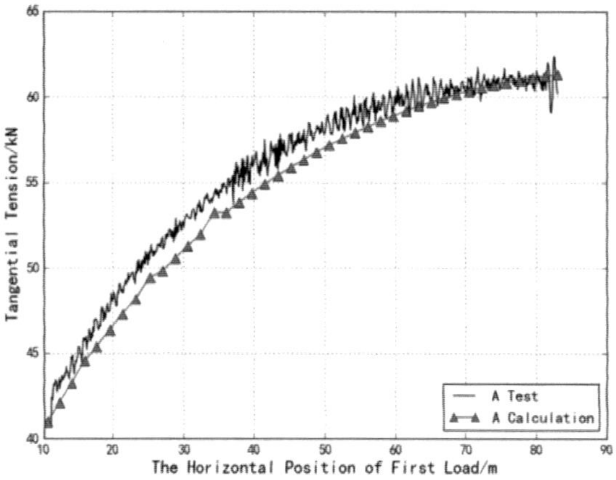

(b) Result of case 2

Figure 4. The results of the tangential tension with the load position in case 1 and case 2

From the figure 4, the curve of calculated tangential tension is consistent with the experimental values, and the maximum error is only 0.16% in case 1, 1.70% in case 2.

Therefore, the accuracy of method is verified which can meet the requirements of engineering applications.

DOI: 10.1051/matecconf/201713602003

6 Conclusions

In this paper, the nonlinear equation set of working cable of single-span circulating ropeway is proposed. The set includes the constant of spans and altitude differences between bracket, the tension constant of the bearing cable and the traction cable beside load, the length constant of the load running space and the bearing cable length.

The calculation method is fast and good in convergence. It is an effective solution for interaction problem of bearing cable and traction cable, which can meet with requirement of the circulating ropeway design.

Acknowledgment

This research was supported by Science and Technology Research Project of State Grid (Research on the calculation method of multi-bearing cable ropeway and the application in UHVDC).

References

1 Zhao Yueyu, Wang Lianhua, etc. Direct Treatment and Discretization of Non-linear Dynamics of Suspension Cable[J]. Acta Mechanica Sinica. 2005, 37(3):329-338.
2 Cheng Daye. Exact Element Method for Analysis of Cable Structures[D]. Beijing: Tsinghua University, 2005.
3 Li Bin, Li Yinghui. Dynamic Modeling and Simulation of Flexible Cable with Large Sag[J]. Applied Mathematics and Mechanics. 2000, 21(6): 640-646.
4 Cheng, S. P. , Perking, N. C. Free Vibration of A Sagged Cable Supporting A Discrete Mass[J]. Journal of the Acoustical Society of America. 1992a, 91(5): 2654-2662.
5. Gavrilov, S.. Non-stationary problems in dynamics of a string on an elastic foundation subjected to a moving load[J]. Journal of Sound and Vibration, 1999, 222(1), 345–361.
6. Di Tomasso, F., Di Egidio, A., Luongo, A., 2003. Modelling and Dynamic Analysis of Strings Carrying A Single Mass. In: Proc. of 16th AIMETA Congress, CD-Rom, Ferrara, Italy (in Italian).
7. Qin Jian, Xia Yongjun. The Matrix Iteration Method for Analysis of Suspension Cable Based on Segmental Catenary Theory [J]. Chinese Journal of Engineering Design, 2013, 20(5): 404~408.

DOI: 10.1051/matecconf/201713602004

Design and Analysis of the Rotatable Balance-arm System in Crane

Jian Qin[1,a], Jiancheng Wan[1], Ming Jiang[1], Yimin Ma[1] and Liang Feng[1]

[1]*China Electric Power Research Institute, 100055, Beijing, China*

Abstract. The merits and demerits of the movable balance-weight schemes in the existing tower crane are summarized and analyzed. The new design of rotatable balance-arm, latticed crane head, platform and rotational upper support is put forward for top of crane SXD50. The rotatable balance-arm is composed of two parallel truss structures, platform, luffing mechanism and balance-weight. The balance-arm can be pulled to rotate around the rotational upper support by luffing mechanism under amplitude control system. The forward torque is equaled to the backward torque from the balance-arm to realize the moment equilibrium of crane body during working. The luffing mechanism of balance-arm on the platform can be as a part of balance-weight. The finite element model of balance-arm and crane head is built and the calculated results prove the rotatable balance-arm system reliable and the structure strength satisfies the requirement of design.

1 Introduce

Tower crane SXD50 has stationary balance-weight installed at the end of balance-arm[1]. No matter how the working range and lifting weight change, the bending moment is constant for the crane body. So the crane body has large cross-section to bear the maximum forward torque caused by load and the maximum backward torque by balance-weight[2].

The movable balance-weight of crane can generate equal backward torque during the operation to decrease the bending moment of crane body.

Currently, there are 4 kinds of movable balance-weight in tower crane[3-6]. (1) The self-adaptation balance-weight realized by crane-slider mechanism. (2) The forward and backward sway balance-weight realized by four bar linkage. (3) The movable balance-weight realized by reciprocating cycle wire rope. (4) Balance-weight pulled by independent tractor system along balance-arm. The four methods mentioned above have various kinds of deficiencies and not adapted for the SXD50 crane.

Among the methods, crane-slider mechanism and four bar linkage need not add auxiliary power to drive. They have the following characteristics: simple structure and clear force-bearing state. But their range of motion is smaller. So, the maximum backward torque supplied can't balance the forward torque generated when loaded fully. In addition, using the two methods above, lifting-arm and balance-arm can't curl in the top of crane and demolish from center after construction, because of the auxiliary structure arrangement, such as link mechanism and linkage device.

In conclusion, the four methods mentioned above have various kinds of deficiencies. So, they are not adapted to the movable balance-weight design of crane SXD50. Therefore, for the crane construction characteristics of transmission line tower erection, developing a new type of movable balance-weight, which not only achieves the predetermined lifting ability, but also reduces the quality of total weight, is an important issue to be solved for the construction of high voltage transmission line.

99i9iIn this paper, a new type of movable balance-weight system is proposed, which is suitable for design of different cranes.

2 The rotatable balance-arm design

To form the design of rotatable balance-arm, the new type and structure are studied.

2.1 The integrated pattern

Based on the characteristics of crane SXD50, a series of new structures are proposed, including rotatable parallelogram balance-arm, latticed head and platform.

Shown as Figure 1, there is small-sized vertical luffing mechanism is arranged on the platform of rotatable parallelogram balance-arm. The balance-weight is set under of platform. The balance-arm is pulled by luffing mechanism under amplitude control system in working, the movement of balance-arm makes the backward torque generated by balance-weight equal to the forward torque by working lifting-arm. This design can realize the goal of reducing the lopsided moment of crane body.

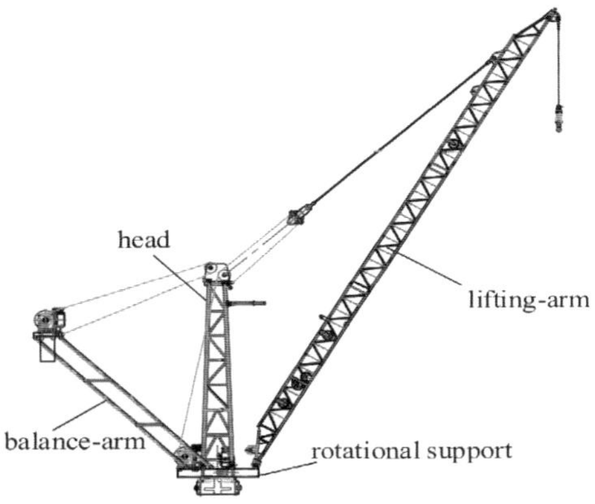

Figure 1. The pattern of rotatable balance-arm system

2.2 Rotatable parallelogram balance-arm

The rotatable balance-arm, which is located on the rotational support, includes two parallel truss structures, four connecting rods and platform. All parts are connected by pins. As shown in Figure 2, the parallelogram structure can remain the platform horizontal and make the luffing mechanism work normally. The minimum work angle of balance-bar is 7°, maximum is 87.5°.

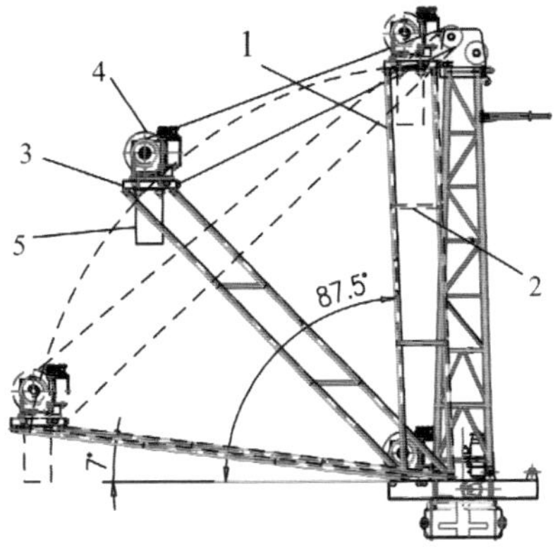

1- Balance-arm, 2-Connect rod, 3-Platform, 4-Luffing mechanism, 5-Balance-weight

Figure 2. Rotatable parallelogram balance-arm

Shown as Figure 3, the upper truss is designed with opening downwards to assure the balance-arm not impact with the crane head. The lower truss has enough blank in the end to make the balance-weight pass through.

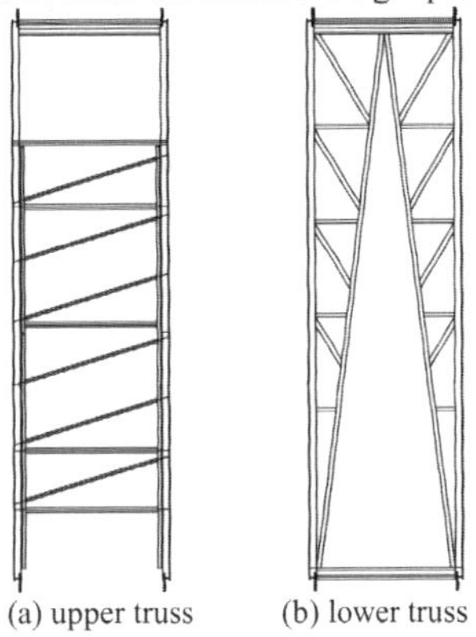

Figure 3. Balance-arm truss

2.3 Platform of balance-arm and balance-weight

 DOI: 10.1051/matecconf/201713602004

The platform (shown as Figure 4) is box-type structure. The luffing mechanism of balance-arm installed on the platform generates the backward torque with weight block under platform together.

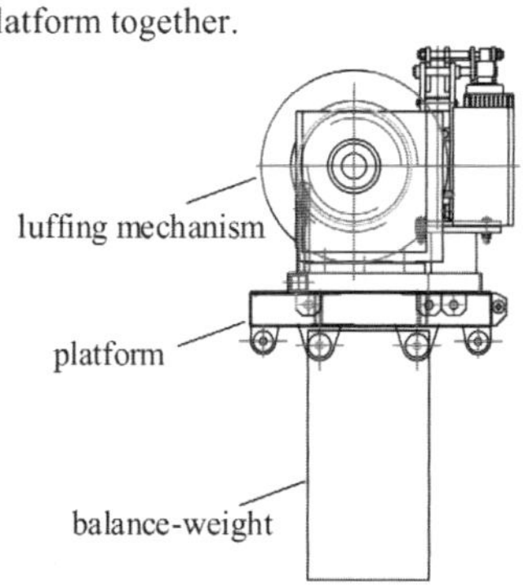

Figure 4. Platform

2.4 Crane head

The head is latticed structure located in the center of rotational support. There is a limit putter on the head to avoid the impact of lifting-arm and balance-arm. The rotational assembly is shown as in Figure 5.

Figure 5. Crane head

2.5 Rotational upper support

The rotational upper support is shown in Figure 6, and the hinged joint of lifting-arm end is arranged on the edge of rotational support. The slewing gear and jib lubbing mechanism is on the support.

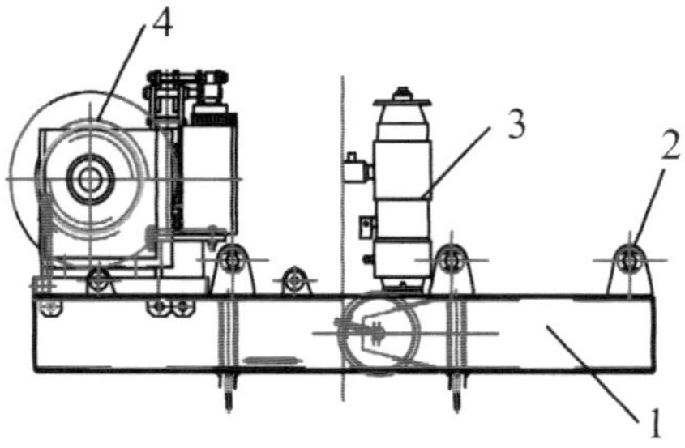

1-Rotational upper support, 2-Hinged joint, 3- Slewing gear, 4-Jib lubbing mechanism

Figure 6. Rotational assembly

3 Calculation of balance-arm structure

To verify that the structure designed meets the strength requirement, the balance-arm and crane head are calculated here. The rotational support is small changed compared with the traditional type, so not check[7].

3.1 Balance-arm

DOI: 10.1051/matecconf/201713602004

The loads on the balance-arm include the weight of counter, platform, luffing mechanism, balance-arm, the wind load and the tension of wire rope. The wind load of balance-arm can be got as:

$$F_w = C_w p_w \omega l h \quad (1)$$

Here C_w is wind coefficient 1.6, p_w is rated wind pressure 250Pa, ω is structure full ratio 1.0, l is length of balance-arm, h is width of balance-arm.

The result of finite element analysis is shown in Figure 7. The maximum equivalent stress of balance-arm is 50MPa, which is much less than the allowable stress $[\sigma]_{Q345B}$=259.40MPa. So the truss structure of balance-arm can meet the strength requirements.

3.2 Tower head

The tower head is mainly loaded with the tension force of lifting-arm and balance-arm. So there are 2 important work cases: 1) 3.125t load (overloading 25%) at 21m and the amplitude of balance-arm is the maximum 8m; 2) 5t load (overloading 25%) at 12.5m and the amplitude of balance-arm is 8m.

The result is shown in Figure 8. The maximum equivalent stress of head is 70.8MPa in case 1, 96.1MPa in case 2, much less than the allowable stress. So the lifting-arm and the whole structure can meet the strength requirements.

Figure 7. The result of balance-arm

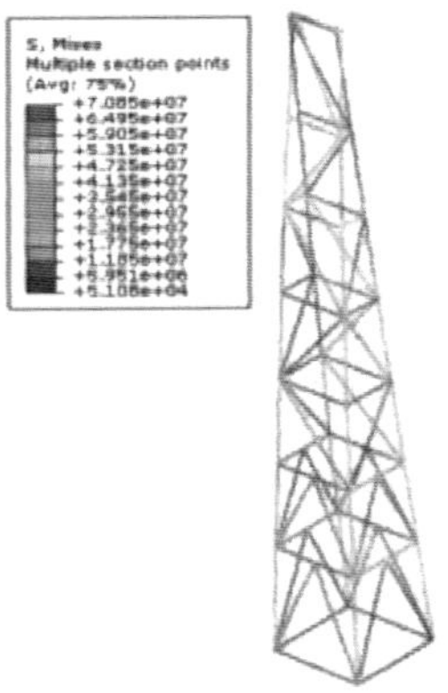

(a) Case 1

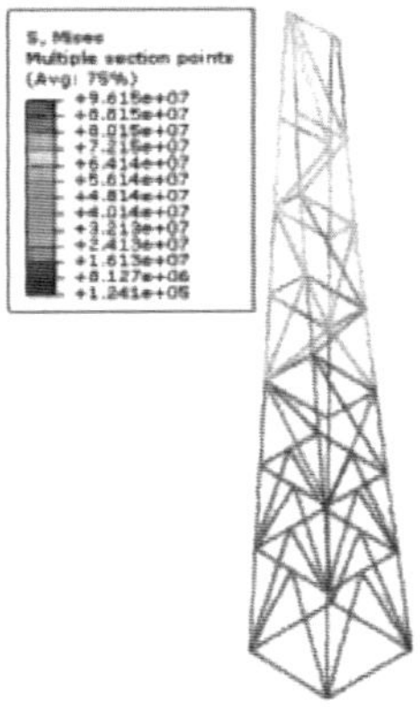

(b) Case 2

Figure 8. The result of head

4. Conclusion

In this paper the new moveable balance-weight system is proposed based on crane SXD50.

Compared with the traditional type, the balance-weight can rotate around platform pulled by the luffing mechanism. The design can realize the moment equilibrium between the forward torque and the backward torque, which not only achieves the predetermined lifting ability, but also improves the force status of crane structure.

All the parts of moveable balance-weight system meet the strength requirements with a big redundancy. It is available to carry out the optimization research to reduce the total weight of crane.

Acknowledgment

This research was supported by Science and Technology Research Project of State Grid (Study on the key design technologies of mobile crane for tower erection in transmission line).

References

1. LI Qinglin. Scheme Selection of Holding Poles Erection for EHV Transmission Towers[J]. Electric Power Construction, 2007, 28(03): 29-33.
2. ZHANG Songhua, YU Qiu-an. Development and Application of LB-3 Crane[J]. Hubei Electric Power, 2010, 34(6): 56-57.
3. ZHANG Ying, YAO Yan-an, ZHA Jianzhong. A Movable-Counterweights Method for Balancing of Mechanisms[J]. JOURNAL OF SHANGHAI JIAOTONG UNIVERSITY, 2006,40(12): 2039-2045.
4. ZHANG Mingqin, SUN Xiaoli, QU Zhenbo, SONG Chuanzeng, DONG Mingxiao. The Self-adaptation Regulating of Crane's Balanced Weight[J]. Machine Design and Research, 2003, 19(4): 44-46.
5. LIU Jiehai. Balance weighed mobilizable device of symmetrical pulling circulation luffing tower crane[J]. CONSTRUCTION MECHANIZATION, 2010(9): 64-66.
6. HOU Baojia, HUANG Qiong, TU Xunian. Optimized design of jib of derrick tower crane[J]. CONSTRUCTION MECHANIZATION, 2006(11): 24- 25.
7. WU Haorong, WANG Xiaoping, GOU Dunpan, LI Xiaoxiao. Optimization Design of Tower Crane Mast Based on ANSYS Zero-order Method[J]. Construction Machinery and Equipment, 2014, 45(2): 42-46.

Design of Resistive Grain Moisture On-line Detector

Chen Chen[1], Chaohua Wu[1,2,a] and Cunrong Li[1]
[1]School of Mechanical and Electrical Engineering, Wuhan University of Technology, Wuhan 430070, China
[2]Hubei Digital Manufacturing Key Laboratory,Wuhan University of Technology,Wuhan 430070,China

Abstract. Grain moisture detection is of great significance to the acquisition, transportation, storage, processing and trade of grain. This paper designs and develops a resistive grain moisture on-line detector, which uses the least squares method for data calibration and stores the calibrated data in the STC15F2K08S2 microcontroller. The grain moisture content is obtained by processing the collected grain resistance and grain temperature. When the grain moisture content exceeds the set value of grain moisture detector, the microcontroller will automatically control the opening of grain drying equipment so as to prevent the grain moisture content from being too high and causing mildew. Through the experimental data analysis and operation test, the detector can accurately detect the moisture content of wheat, barley, rapeseed, rice and other grain and automatically control the drying equipment, and it has the advantages of fast response, high accuracy and strong practicality, so it can meet the national grain testing accuracy requirements and automatic drying requirements.

1 Introduction

The survival of mankind is based on grain, and grain plays an important role in national economic construction and all walks of life. In order to ensure the quality of grain, it is necessary to accurately measure the moisture content of grain before grain storage. However, due to many factors, the traditional method is difficult to achieve real-time detection of grain moisture in the process of grain drying, as well as intelligent control of drying equipment[1]. Therefore, the modern science and technology theory are applied to achieve moisture on-line accurate detection and intelligent control of drying equipment in the process of grain drying, which has a significant role in improving grain quality and economic efficiency.

According to the different principles of grain moisture detection, the detection method can be divided into the following: drying method, chemical method, electrical measurement method, near infrared method, microwave method [2,3]. Drying method and chemical method measure the grain moisture value through the heating or chemical reaction and other means. This measurement method has the advantages of accuracy and reliability, but it wastes time. Generally, it is only suitable for professionals to operate. The electrical measurement method includes the resistance method and the capacitance method, which measure the grain moisture value by measuring the dielectric constant of the grain[4,5]. This measurement method has the advantages of simple principle, convenient operation and low cost, but the capacitance method is easy to be influenced by temperature and other factors[6].

Near infrared method and microwave method have the advantages of high accuracy and fast response,[a]but their structure is complex and they are expensive. They are rarely used at present.

Therefore, this paper designs a resistive grain moisture on-line detector, which uses the resistance method to detect the grain moisture in the grain drying equipment in real time, and uses the single chip microcomputer to automatically control the operation of the grain drying equipment. The detector can prevent the grain moisture content from being too high and producing mold. And it saves a lot of manpower and material resources. The detector structure is simple and the measurement results are accurate. And it can enable the production process to be automated and intelligent.

2 Detection principle of resistive grain moisture on-line detector

Experimental studies have shown that when the grain moisture content is between 9% and 20%, the grain moisture content is an approximate linear relationship with the logarithm of grain resistance, and when the moisture content increases, the grain resistance will rise from tens of megohm to less than one thousand ohm. At the same time, there is a linear relationship between grain resistance and grain temperature. When the temperature is between -10°C and + 50°C, the grain moisture content will be reduced by 0.1% for every 1°C rise in temperature[7]. Therefore, grain moisture content can be obtained by measuring the grain resistance and grain temperature in grain drying equipment.

[a] Corresponding author : wuchaohuavip@163.com

Resistive grain moisture on-line detector uses constant current source circuit to measure grain resistance and temperature compensation circuit to measure the grain temperature in the drying equipment. Constant current source circuit can make the power supply output voltage constant and the current flowing through the grain resistance constant and large. So when the voltage flowing through the grain resistance is measured, the grain resistance value can be determined, and then grain moisture content can be obtained.

3 Composition and principle of resistive grain moisture on-line detector

3.1 The overall structure

Figure 1 shows the structure of the resistive grain moisture on-line detector, which is composed of the lower computer and the host computer. The lower computer is mainly composed of electrode roller and shell. The electrode roller is a rotatable crushing electrode and is mainly used to crush the grain between the gap of the electrode roll to collect grain resistance. The host computer is mainly composed of knobs, buttons, digital tube and shell. The knobs and buttons are mainly used to select the types of grain to be tested and the detection methods. The digital tube is mainly used to display the grain moisture value.

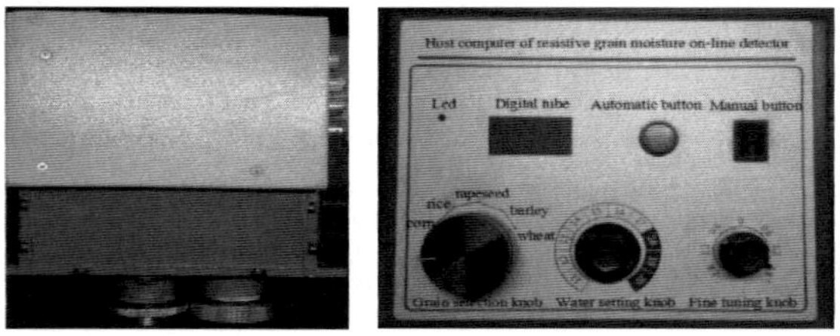

Lower computer Host computer

Figure 1. The structure of the resistive grain moisture on-line detector.

3.2 Working principle

The lower computer is installed at the hopper of the grain drying equipment, its internal motor drives the electrode roller to rotate, crushing the grain particles between the gap of the electrode roll. At the time, the constant current source circuit will collect the potential difference on the grain resistance, the potential difference is amplified by the signal amplification circuit and transmitted to the single chip in the lower computer, single chip gets the grain moisture value by A/D conversion. The moisture value is corrected by the temperature compensation circuit and transmitted to the single chip in the host computer for display. At this time, the single chip in the host computer controls the start and stop of the motor and the drying equipment by comparing the measured grain moisture value with the set moisture value, the working principle flow chart is shown in Figure 2.

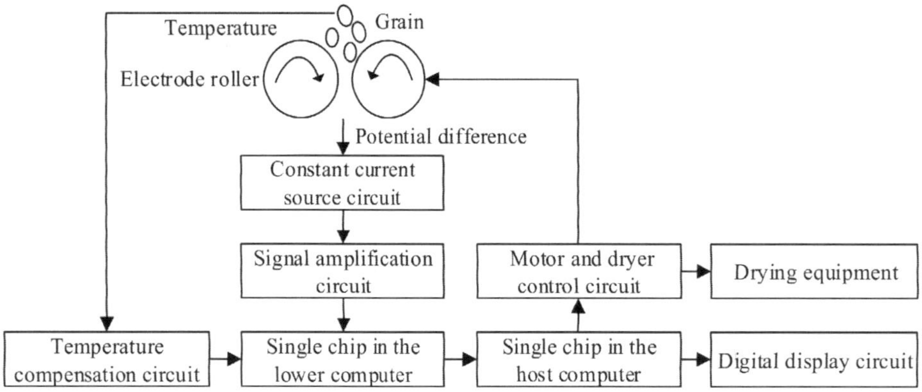

Figure 2. The working principle flow chart.

3.3 Hardware circuit composition

The hardware circuit of the resistive grain moisture on-line detector mainly includes the lower computer circuit and the host computer circuit.The lower computer circuit is mainly composed of power supply circuit, constant current source circuit, signal amplification circuit, temperature compensation circuit, and single chip system. The host computer circuit is mainly composed of power supply circuit, digital display circuit, motor and dryer control circuit, control knob and button circuit, and single chip system, the hardware circuit block diagram is shown in Figure 3.

DOI: 10.1051/matecconf/201713602005

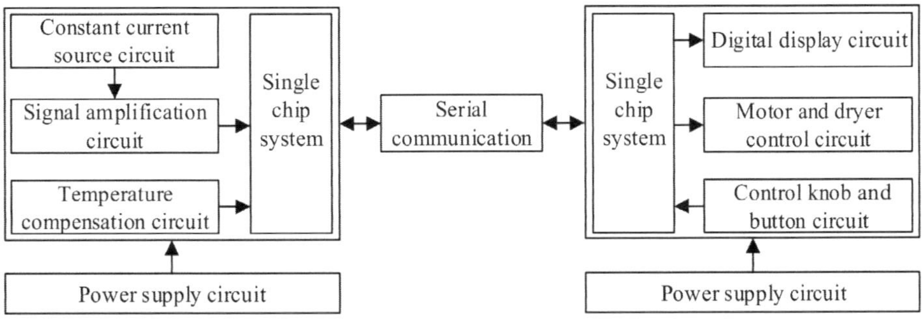

Figure 3. The hardware circuit block diagram.

3.4 Software design

Figure 4 shows the main program flow diagram of the resistive grain moisture on-line detector, which is designed in a modular way. And according to the needs of the main program, the system will be initialized and subroutine calls.

The work process of the resistive grain moisture on-line detector is as follows: after the resistive grain moisture on-line detector power is running, first rotate the "grain selection" knob on the top panel to select the grain species to be dried in the drying equipment, at this time, wheat, barley, rapeseed, rice, and corn can be selected. Then adjust the "water setting" knob, the set value of the grain moisture detector will be displayed on the digital tube. Finally select the measurement method, at this time, the motor will drive the electrode roller to rotate, the drying equipment will also begin to work, the digital tube begins to measure the number of the extruded grain, when the number reaches 100 and the measurement value of the grain moisture is less than the set value, the motor and drying equipment will stop working. At this time, the value of the digital tube is the grain moisture value, otherwise, the motor and drying equipment will continue working.

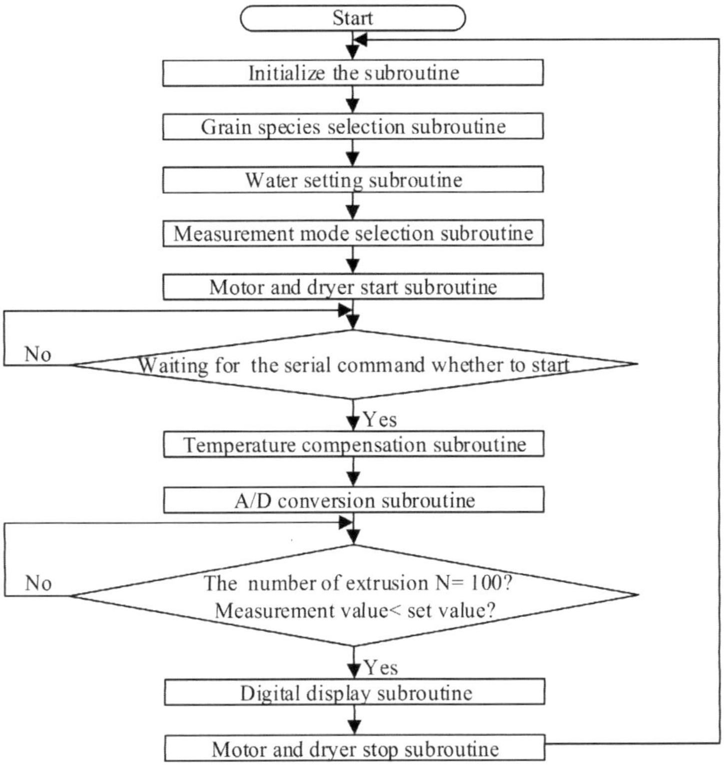

Figure 4. The main program flow diagram of the resistive grain moisture on-line detector.

4 Calibration and error test of resistive grain moisture on-line detector

 DOI: 10.1051/matecconf/201713602005

4.1 Calibration experiment

Under the condition of room temperature, 20 kinds of wheat samples with the same variety and different humidity were selected, and the actual water value of wheat was measured by Qifeng QLY-T grain moisture analyzer. The potential difference of the output end of the signal amplification circuit was measured by Fluke 15B+ high precision digital multimeter respectively. The signal amplification circuit amplifies the potential difference on the wheat resistance by a factor of 50. The results of the measurement data are shown in Table 1.

Table 1. Measurement data of actual water value and amplified potential difference.

Sample numer	Water value W (%)	Potential difference lgu (mV)	Sample numer	Water value W (%)	Potential difference lgu (mV)
1	9.8	1.342	2	10.3	1.362
3	10.5	1.378	4	10.9	1.396
5	11.5	1.439	6	11.8	1.467
7	12.4	1.472	8	12.7	1.497
9	13.4	1.526	10	14.3	1.583
11	15.0	1.618	12	15.2	1.643
13	15.8	1.668	14	16.7	1.701
15	17.2	1.721	16	17.6	1.740
17	18.1	1.761	18	18.7	1.808
19	19.5	1.845	20	19.8	1.890

The data measured in Table 1 are fitted using the least squares method in matlab. The fitting results are shown in Figure 5. With the increase of water content of wheat, the logarithm of the potential difference will also increase. When the water content of wheat is between 9.8% and 19.8%, the logarithm of the potential difference on wheat resistance is approximately linear with the moisture content of wheat.The equation is lgu = 0.0523 W +0.831. Similarly, the calibration of barley, rapeseed, rice and corn can be carried out.

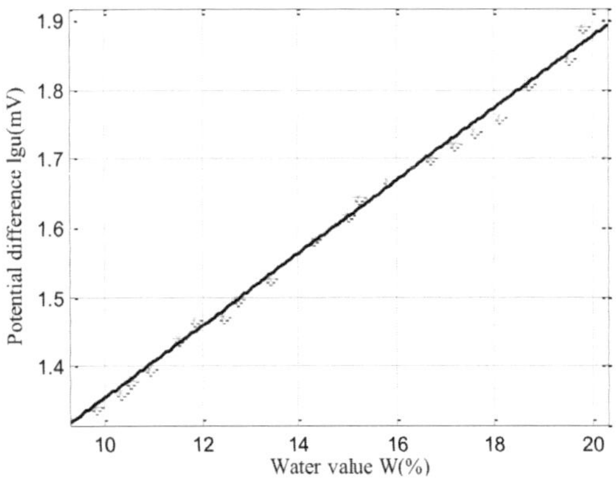

Figure 5. The fitting result of actual water value and amplified potential difference.

4.2 Error test

Table 2 shows the results of the moisture measurements of wheat which are measured by the self-designed resistive grain moisture on-line detector and the actual moisture value of wheat which are measured by Jufeng QLY-T grain moisture analyzer. As can be seen from the data in the table 2, the minimum resolution of the moisture detector reached 0.5% of the national standard, and the measurement results are accurate[8].

Table 2. Comparison of moisture measurements and actual moisture value of the wheat.

Sample numer	Measurements(%)	Actual value (%)	Contrast error
1	9.88	9.91	0.03
2	11.35	11.34	0.01
3	13.50	13.49	0.01
4	14.32	14.32	0
5	15.76	15.79	0.03
6	16.20	16.18	0.02
7	16.72	16.68	0.04
8	17.16	17.16	0
9	18.12	18.14	0.02

 DOI: 10.1051/matecconf/201713602005

10	19.37	19.36	0.01

5 Conclusion

This paper analyzes the principle of resistive grain moisture on-line detector and puts forward the method of measuring grain resistance by constant current source circuit. On basis of this, hardware design, software design and instrument calibration are completed. Through the experimental test, the resistive grain moisture on-line detector can accurately detect grain moisture content and control grain drying equipment.And it has advantages of automatic operation, fast response, high precision, low cost, strong practicality. So its application prospects are very broad.

6 Acknowledgement

This work was supported by the Fundamental Research Funds for the Central universities (WUT: 2017-IVA-117).

References

1. GL. Zhang. Jilin Agricultural University (2014).
2. AC. Li, ZQ. Wang, L. Li, F. Gao. Agriculture Network Information, 2, 114-116 (2009).
3. HX. Zhang. Shandong University of Technology (2012).
4. BF. Zhai, HL. Guo, H. Xu. Journal of Shenyang University of Technology, 23, 413-416 (2011).
5. BF. Zhai, Y. Bai. Journal of Liaoning Institute of Technology, 23, 34-39 (2003).
6. J. Sun, ZM. Zhou, HJ. Tang. Food Storage, 36, 46-48 (2007).
7. Z. Liu, ZD. Wu, et al. Mathematical Problems in Engineering, 2015, 1-10 (2015).
8. Y. Jing, L. Ding. Journal of Shenyang Aerospace University, 28, 51-54 (2011).

MATEC Web of Conferences **136**, 02006 (2017) DOI: 10.1051/matecconf/201713602006
D2ME 2017

The analysis of surface topography during turning of Waspaloy with the application of response surface method

Damian Przestacki[1,a], Tadeusz Chwalczuk[1]

[1]*Poznan University of Technology, Institute of Mechanical Technology, Piotrowo 3, Poznan 60-965, Poland*

Abstract. This paper presents the analysis of surface topography during longitudinal turning of Waspaloy. The primary objective of the work was to determine the optimal machining conditions enabling the minimization of surface roughness parameters. Experimental studies were performed with the application of fractional plan (with 1 block and 9 systems). The carried out experiment included the measurements of surface topographies formed after turning with variable cutting conditions (cutting speed v_c, cutting depth a_p and feed f). Subsequently the measured quantities have been applied to the optimization procedure based on the application of response surface method (RSM). In order to obtain the optimal cutting parameters, the maximization of total utility function was carried out. Results revealed that the minimal surface roughness parameters values during turning of Waspaloy with the cubic boron nitride (CBN) inserts can be obtained for the: f=0.05 mm/rev, a_p =0.23 mm and v_c =452 m/min.

1 Introduction

Advanced aerospace nickel-based alloys are designed to work in very heavy loads at the elevated temperatures. Nevertheless, these super alloys have low machinability because of their properties focused on improved resistance to hot corrosion, oxidation, creep strain and dwell crack growth, higher stability of the microstructure and yield stress at the high temperature [1]. According to Ezugwu *et al.* [2], the nickel-based superalloys can be characterized by the three distinct groups, namely: nickel–copper, nickel–chromium and nickel-molybdenum–chromium.

Waspaloy is a nickel-chromium alloy with very high strength at the elevated temperatures and excellent corrosion resistance. Primarily, this material is applied in aerospace and automotive industries due to its high corrosion resistance and strength at high temperatures. Nevertheless, these unique exploitation properties significantly hinder Waspaloy's machinability. The primary problems occurring during machining of Waspaloy are intense tool wear [3, 4] and high surface roughness [5]. According to Takur and Gangopadhyay [6], during machining of nickel-based alloys, surface roughness is strongly affected by the three major factors: the initiation of workpiece material's plastic side flow, the appearance of built up edge, and the progression of tool wear. In addition, the machined surface profile can contain also the defects including feed marks, laps and grooves [7]. In order to improve the machinability of hard-to-cut materials, the modern machining trends can be applied. These methods include the high speed machining (HSM) [8-10], the laser assisted machining (LAM) [11-13], as well as the machining with the application of minimum quantity cooling lubrication

[a] Corresponding author : damian.przestacki@put.poznan.pl

(MQCL) [14, 15]. Another method which can be successfully applied to select the effective machining conditions, enabling the improvement of process efficiency or quality, is the optimization. The machining process' optimization objectives are usually focused on the minimization of surface roughness [16], cutting forces [17], elongation of tool life [18], reduction of tool wear [19] and machining vibrations [20]. The popular optimization approaches are mainly based on the Taguchi method and determination of signal-to-noise ratio, grey relational analysis (GRA), as well as the response surface methodology (RSM). Nevertheless, the researches focused on the optimization of Waspaloy machining process have not been reported in literature.

The literature survey reveals that despite many researches regarding the nickel-based superalloys, the machinability of Waspaloy is insufficiently recognized. Therefore, this study is focused on the evaluation of surface topography and roughness parameters during turning of Waspaloy material. The carried out research involves the optimization of cutting parameters (feed per revolution f, cutting depth, a_p and cutting speed v_c), allowing the improvement of surface finish obtained after longitudinal turning of Waspaloy.

2 Experimental details

2.1 Work and tool materials

The Waspaloy super alloy was used as the workpiece material. Its chemical composition is depicted in Table 1 [5]. The workpieces were provided as a cylinders with a diameter of 80 mm. The longitudinal turning tests were conducted with the application of the cubic boron nitride CBN7015 inserts with the VBGW 160408 S01020F symbol and the following geometry: tool major cutting edge angle $\kappa_r = 72.5°$, flank angle in the feed motion direction $\alpha_f = 5°$, and tool radius $r_\varepsilon = 0.8$ mm.

TABLE 1. Chemical composition of waspaloy [%]

Element	Content
C	0.036
S	0.0004
Mn	0.03
Si	0.05
Cr	19.87
Mo	4.27
Co	13.26
Ti	3.08
Al	1.35
B	0.007
Zr	0.06
Fe	0.09
Cu	0.01
Ni	57.32
P	0.004

2.2 Machining tests

The experiments (see – Tab. 2) involved the longitudinal turning without fluids, carried out on precise TUR 560E lathe with variable cutting speed v_c, cutting depth a_p and feed f. The experiment was designed with the application of fractional plan (with 1 block and 9 systems) in order to reduce the time and costs of carried out tests.

Table 2. Design of experiment

No	Level of variability			Feed f [mm/rev]	Cutting speed v_c [m/min]	Depth of cut a_p [mm]
1	0	1	-1	0.040	226	0.30
2	-1	-1	-1	0.040	89	0.10
3	-1	0	1	0.112	89	0.20
4	1	1	1	0.112	452	0.30
5	1	-1	0	0.077	452	0.10
6	-1	1	0	0.077	89	0.30
7	1	0	-1	0.040	452	0.20
8	0	0	0	0.077	226	0.20
9	0	-1	-1	0.112	226	0.10

2.3 Inspection of topography

The surface profile meter, type Veeco NT 1100 has been applied to the evaluation of surface topography. The measurements were made with 5.1-fold magnification. The measurements were repeated 3 times for the each investigated machined surface. The acquired signals were processed in Veeco Vision 32 software. As a consequence, the average surface roughness *Ra* and surface roughness height *Rz* were determined.

2.4 Optimization of Waspaloy turning process

The multi-criteria optimization procedure was applied in the research. This optimization was based on the transformation of the approximated output factors' values into the singular value of the resultant total utility U_{tot}. Thus, the problem of simultaneous optimization of many output factors was simplified to the determination of one output parameter, which maximizes total response utility. In the carried out research, the *Ra* and *Rz* surface roughness parameters were applied as an optimization criteria. In the first part, the turning tests conducted on the basis of the fractional experimental design were carried out. The obtained tests' results were the starting point to the definition of response utility profile, formulated for the investigated output factors (*Ra*, *Rz*). During this optimization stage, all output factors' values were assigned to the partial utilities, which values were contained in the range: <0, 1>. After the definition of response utility profiles, the optimal solution's searching procedure was applied. This procedure included the mesh definition of the input variables. Subsequently, the output factors' predictive values were calculated for the all mesh points and as a result, the partial utility values were received. In order to determine the total utility value, the mean geometric values of partial utilities were calculated in the each mesh point. The next step involved the formulation of the total utility model, with the application of spline function. After the generation of this model, the optimal solution searching algorithm was adopted. The optimal solution determines the values of input variables, corresponding to maximal value of the total utility. In order to achieve it, the simplex method implemented in the **StatSoft Statistica 13.1** software was applied. Ultimately, the optimal input parameters' values were received.

3 Results and discussion

The surface topographies, obtained during turning in the range of the lowest (experiment 2) and the highest (experiment 4) levels of selected cutting parameters are presented in figure 1. It can be observed that for the both investigated cases (Fig. 1a, Fig. 1b), the feed marks on surface topographies are seen, resulting from the kinematic-geometric model [21]. Furthermore, in case of the lowest selected cutting parameters, the surface topography is less regular and characterized by the appearance of burrs located mainly at the peaks of irregularities. This phenomenon can be attributed to the high ploughing action, resulting from the intense elastic-plastic deformations of workpiece occurring at the low chip thicknesses (correlated with feed and depth of cut values).

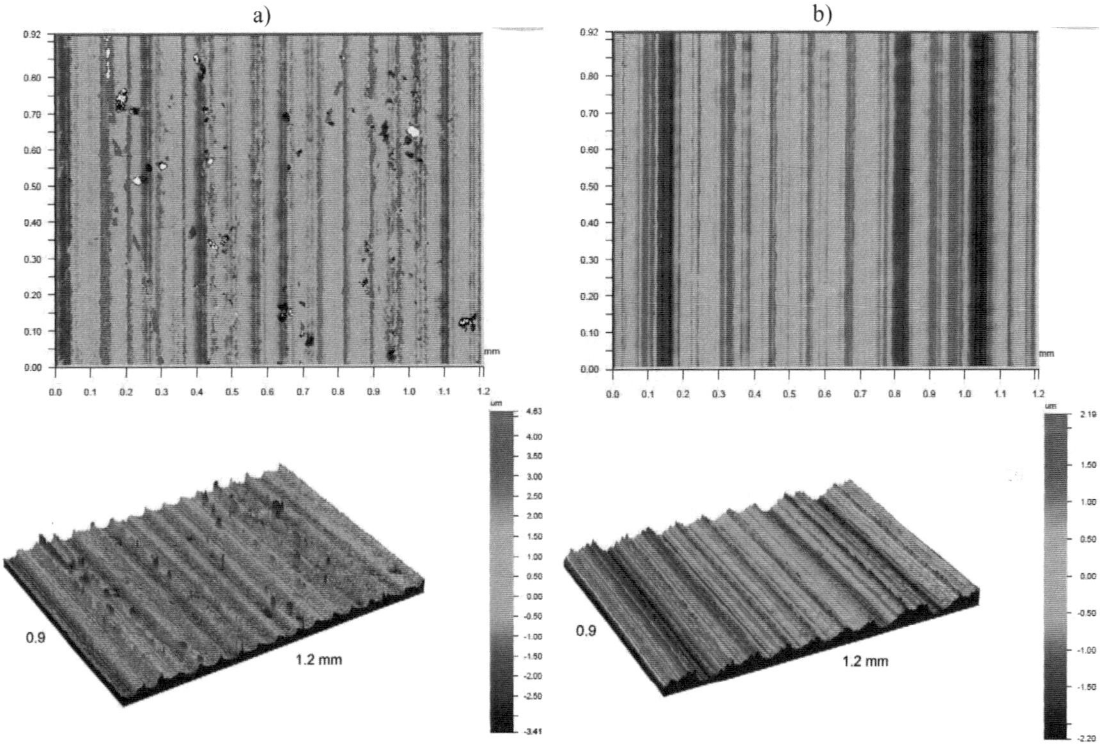

Figure 1. Surface topographies of Waspaloy samples obtained during turning with: a) v_c = 89 m/min, a_p = 0.1 mm; f = 0.04 mm/tooth; b) v_c = 452 m/min, a_p = 0.3 mm; f = 0.112 mm/tooth.

The analysis of surface roughness *Ra* response surfaces, formulated for the investigated cutting parameters (Fig. 2), shows similar qualitative relations independently on selected cutting depth. From these figures it can be seen, that surface roughness *Ra* values have an increasing trend with the growth of feed per revolution. This is a typical relation resulting from the traditional kinematic-geometric surface roughness model. In case of cutting speed, the highest surface roughness values were obtained for the moderate v_c values, independently on selected feed per revolution. Figures 2a-2c show also that during turning with the lowest selected cutting depth values, the surface roughness *Ra* parameters are the highest. This relation can be explained by the occurrence of burrs, located at the irregularity peaks (see – Fig. 1a), which can be caused by the ploughing phenomena occurring at the low chip thicknesses.

The resultant total utility function for the cutting speed v_c, depth of cut a_p, and feed per revolution f is presented in Figure 3. The maximal value of total utility function U_{tot} is achieved with feed per revolution f = 0.05 mm/rev, cutting depth a_p =0.23 mm and cutting speed v_c =452 m/min. The application of these optimal cutting parameters gave surface roughness parameters Ra<0.5 μm and Rz<3.5 μm.

It can be also concluded (Figure 3) that the optimum cutting parameters are for the highest value of cutting speed. The selection of the higher cutting speeds during machining of Waspaloy can result in the growth of tool wear intensity. Therefore, the further studies, focused on the analysis of tool wear in function of cutting speed are required to verify the validity of the carried out optimization.

 DOI: 10.1051/matecconf/201713602006

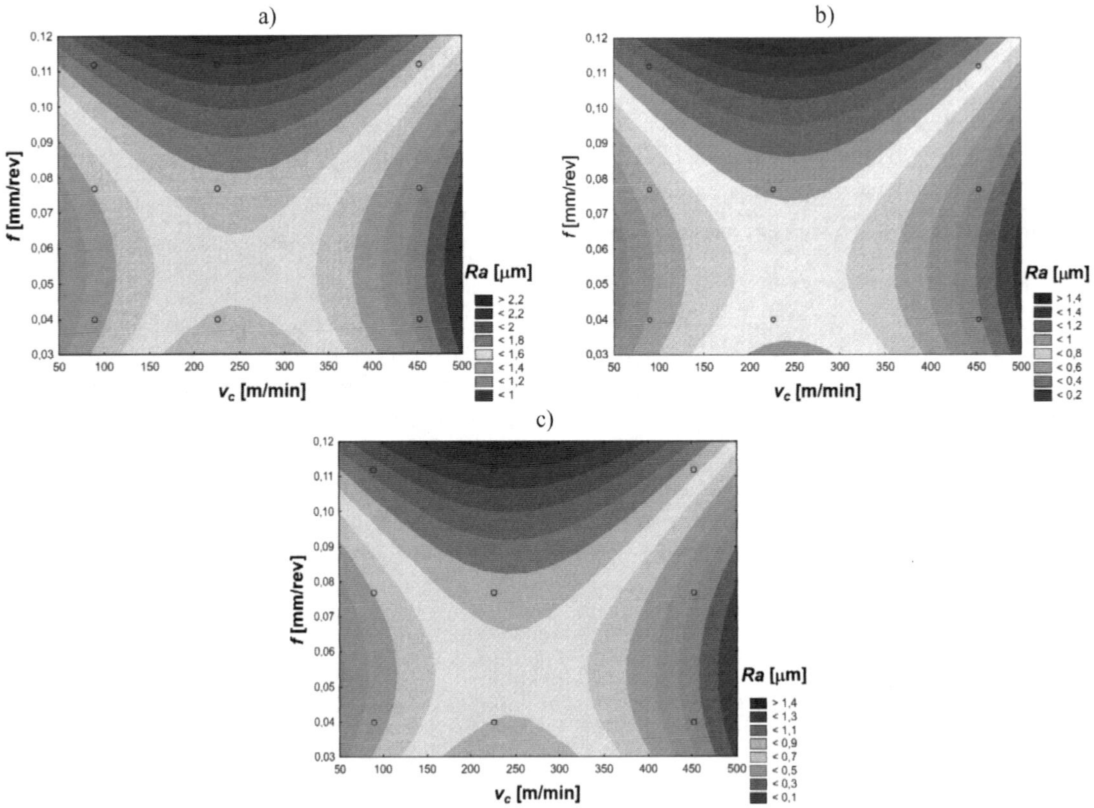

Figure 2. Surface roughness *Ra* parameters of Waspaloy samples obtained during turning with: a) a_p = 0.1 mm; b) a_p = 0.2 mm; c) a_p = 0.3 mm.

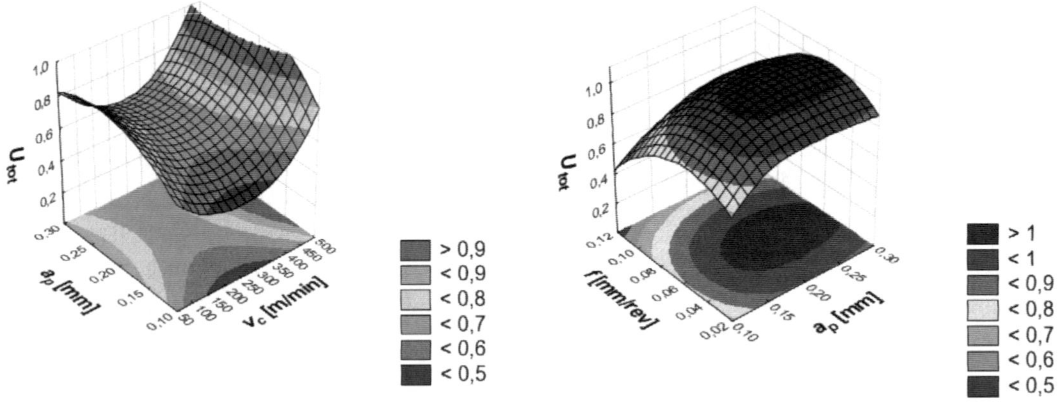

Figure 3. Total utility function for the cutting speed v_c, depth of cut a_p, and feed per revolution *f*.

4 Conclusions

On the basis of conducted research, focused on the analysis of surface topography during longitudinal turning of Waspaloy, the following conclusions have been formulated:

- the feed marks on surface topographies are seen, independently on the selected cutting parameters. In addition, in case of the lowest tested cutting parameters, the surface topography is less regular and characterized by the appearance of burrs located mainly at the peaks of irregularities;
- during turning with the lowest investigated cutting depth values, the surface roughness *Ra* values are the highest. This relation can be attributed to ploughing phenomena occurring at the low chip

thicknesses;

- based on the conducted optimization, the following optimal turning parameters were obtained: f = 0.05 mm/rev, a_p =0.23 mm, v_c =452 m/min.

Acknowledgment

The presented research results, executed under the domestic project PBS of No 244445, were funded with grants for education allocated by the National Centre for Research and Development.

References

1. G. Kappmeyer, C. Hubiga, M. Hardy, M. Witty, M. Busch, Procedia CIRP **1**
2. E.O. Ezugwu, J. Bonney, Y. Yamane, J. Mater. Process. Technol. **134** (2003)
3. S. Olovsjö, L. Nyborg, Wear **282– 283** (2012)
4. R. Polvorosa, A. Suárez, L.N. López de Lacalle, I. Cerrillo, A. Wretland, F. Veiga, Journal of Manufacturing Processes **26** (2017)
5. A.R. Motorcu, A. Kus, I. Durgun, Measurement **58** (2014)
6. A. Thakur, S. Gangopadhyay, International Journal of Machine Tools & Manufacture **100** (2016)
7. A.R.C. Sharman, J.J. Hughes, K. Ridgway, Mach. Sci. Technol. **8**(3) (2004)
8. S. Wojciechowski, P. Twardowski, M. Pelic, R.W. Maruda, S. Barrans, G. Krolczyk, Precision Engineering **46** (2016)
9. S. Wojciechowski, P. Twardowski, M. Pelic, Procedia CIRP **14** (2014)
10. S. Wojciechowski, R.W. Maruda, P. Nieslony, G.M. Krolczyk, International Journal of Mechanical Sciences **119** (2016)
11. D. Przestacki, T. Chwalczuk, S. Wojciechowski, Int J Adv Manuf Technol, DOI 10.1007/s00170-017-0035-5 (2017)
12. D. Przestacki, P. Szymanski, S. Wojciechowski, Composites Part A: Applied Science and Manufacturing **91** 1 (2016)
13. D. Przestacki, L. Marciniak-Podsadna, R. Majchrowski, Applied Surface Science, **388**, part A, (2016)
14. R. W. Maruda, G. M. Krolczyk, M. Michalski, P. Nieslony, S. Wojciechowski, Journal of Materials Engineering and Performance **26** (2017)
15. R.W. Maruda, G.M. Krolczyk, P. Nieslony, S. Wojciechowski, M. Michalski, S. Legutko, Journal of Manufacturing Processes, **24**, 1 (2016)
16. T. Kivak, Measurement **50** (2014)
17. S. Karabulut, Measurement **66** (2015)
18. G.M. Krolczyk, P. Niesłony, S. Legutko, Archives of Civil and Mechanical Engineering **15** (2) (2015)
19. X. Wang, I. S. Jawahir, International Journal of Production Research **43** (2005)
20. M. Subramanian, M. Sakthivel, K. Sooryaprakash, R. Sudhakaran, Measurement **46** (2013)
21. D. Przestacki, M. Jankowiak, Journal of Physics: Conference Series **483** (2014)

Investigating the Impact of Road Condition Complexity on Driving Workload Based on Subjective Measurement using NASA TLX

Sugiono Sugiono[1,a], Denny Widhayanuriyawan[2], and Debrina P. Andriani[1]

[1] *Industrial Engineering Dept., Universitas Brawijaya (UB), Indonesia*
[2] *Mechanical Engineering Dept., Universitas Brawijaya (UB), Indonesia*

Abstract. Prior researchers indicate that mental load is one of the most important contributors to a traffic accident. The aim of the paper is to investigate the impact and the correlation of road condition and driving experience on driver's mental workload. The driving test consists of 3 road complicity situation (urban road, highway, rural road) with 26 drivers with average 21 years old in different experience level (average 4.08 years' experience). NASA TLX questioner is used as subjective driver's mental load measurement with three dimensions relate to the demands imposed on the subject (Mental, Physical and Temporal Demands) and three to the interaction of a subject with the task (Effort, Frustration, and Performance). There are 3 cameras placed on the left side, right side and front car to identify the road condition. According to experiment, it was found that drivers felt that frustration level, business, and mental-demand factors dominate the impact on high-level workload (96.15%). Highway road conditions provide an average overall workload score of 62 (OWS) which was better compared to city road (OWS = 69) and rural road (OWS = 66). Based on street complexity, it is necessary to improve road conditions that resemble highway road by reducing potential hazard.

Keywords: *Mental driving workload, NASA TLX, Overall workload score (OWS)*

I Introduction

Driving is main human activity which is conducted to support daily activities and mobility in the shortest time possible. This initiated large government investment in providing safe and convenient transportation system. The driving complexity causes road accidents, not only due to physical factors such as cars and road conditions but also due to human factors themselves such as human error and excessive stress levels. Fastenmeir and Gstalter [1] research define that the complexity of driving depends on elements such as road design (highway, rural road, city road), road layout (straight, curved, inclined, junction) and traffic flow (high density). The study of the situation and the pattern of the relationship between drivers with cars and driving environments in order to reduce or avoid possible risk. The driver's ability to meet driving needs becomes an important capital in realizing safety driving [2]. The ability to adapt quickly, such as reducing speed, changing the direction of the vehicle is very much dependent on driving experience and mental condition (load) of the driver.

Schlegel [3] defines driving workload as the total amount of resources or driver's ability to achieve a driving task. Further driving workload reflects the driving need to achieve performance tasks that depend on individual drivers and skills. Driving workload depends on the task (demand, complexity, and difficulties), drivers (age, fatigue, abilities, experience, etc.) and environment (dynamic/changing, familiar, etc.). Total driving workload consists of every aspect relating to driving task and non-driving task [4]. Driving tasks consist of street monitors making lateral/longitudinal action by using main control (e.g. steering and pedals) and manipulating other controls related to driving safety (e.g. windshield) and displays. Non-driving tasks such as operating the secondary interface (e.g. AC, audio/video, reading maps, phone, drink, etc.). Both of the above driving tasks require the mental and physical actions. Mental actions include acceptance, information processing and decision making involving human five senses and memory systems. Physical actions involve providing energy to move human organs such as hands, feet, head, and eyes to drive. Driving situations have a major impact on the mental workload manifested in the relationship pattern between task demand and driving performance [5]. The complexity of the environment strongly affects drivers to respond and produce good performances, for example when driving around the market or urban streets with a huge number of passersby, the driver is required to immediately process all the information appropriately. Events like this can cause overload to keep driving performance. Therefore takes the correct strategy and qualified physical condition.

An understanding of mental workload and physical workload on drivers is essential to building a powerful Driving alert system. Shih Nan Lu et al. explains that the Industrial Technology Research Institute (ITRE) has also used the ultra-wide bandwidth (UWB) technique to integrate low power pulsed electromagnetic (EM) waves to precisely measure the driver's physiological signals such as his heartbeat, respiration, etc. [6]. ITRE has developed a driving safety system focusing not only on the physical technology of the vehicle but also considering the human. Furthermore, the project developed will create a driving safety system that involves a mental load that will be measured in a subjective offline with National Aeronautics and Space Administration Task

[a] Corresponding author: sugiono_ub@ub.ac.id

Load Index (NASA TLX) and objective (heart rate, EEG). A driving safety system created will record the initial measurement values of driving and analyse other information (bedtime, physical complaints experienced) to establish the maximum allowable minimum and allowable limits. At the time of driving, if the security system receives the insert value exceeds the maximum limit then the alerting system will provide the mark in the form of sound and visual through the mobile phone application. Therefore the driver will adjust their driving which ultimately reduces the risk of accidents. As an initial stage of research, this paper will exhibit the measurement of subjective mental burden using NASA TLX on the diverse trajectory (market/city road, highway, rural road) and diverse driving experiences.

2 Material and Methodology

Basically, human activity can be differentiated in physical work (muscle) and mental work (brain). Physical and mental activity will lead to the emergence of workload. The workload is the difference between a worker's ability and job demands (Meshkati & Hancock, 1988). Should the ability of the workers is higher than the demands of the work it will cause a sense of boredom and vice versa, if the ability of workers is lower than the demands of the work it will cause excessive fatigue that causes stress work on employees and cause frequent occurrence of human error that led to work accidents.

This study focused on measuring the mental workload of drivers on the street environment, driving tasks, and drivers personality. The method used in this study is NASA-TLX. This method can analyse the mental workload of a driver. The basic reason for selecting the NASA TLX method is the ease of measurement implementation [7]. The NASA - TLX method was developed by Sandra G. of the NASA research center and Lowell E. Staveland of the San Jose State University in 1981. NASA TLX is generally composed of 2 main parts of rating and weighting. The rating was obtained from the respondents on what was felt during the completion of the driving assignment with a rating scale of 0 to 100. This rating factor was developed based on the emergence of subjective measurement needs consisting of 6 scales: Mental demand (MD), Physical demand (PD), Temporal demand (TD), Performance (P), Frustration (F), Effort (EF). Weighting is calculated on the basis of pairing options deemed most relevant to the workload experienced. Weighting is calculated by the tally of 15 pairwise combination options of the mental load factor. Weighting had values between 0 and 5. Overall workload score (OWS) is calculated by multiplying the rating by weighting the subjective measuring factor on each respondent.

This research was developed using field experiment utilizing similar car facility. Subjective mental workload measurement process was conducted in several streets at Malang and its surrounding city, in East Java, Indonesia. The roads involved were highway, rural road, city road/market during busy working hours. The driver after passing the predetermined path was to immediately fill out the NASA TLX questionnaire which corresponds to what is perceived from the 6-dimensional measurement assessment. There are 26 driver samples performed by driving performance measurement with varied driving period experience. Respondent drivers' age was 21 years old on average, with an average driving experience of 4.08 years. Based on Ambarwati et al. [8], drivers in Indonesia experiencing a number of accidents were teenagers aged between 18-24 years old. Questionnaires were given to 26 drivers after passing through the streets as much as 3 repetitions targeted such as market/city road, rural road, and highway.

Steps taken in measuring mental workload of drivers utilizing NASA TLX methods which passed through various conditions and street complexity can be described as follows:

1. Preparation of data collection equipment: questionnaire, camera, etc.
2. Set up experiments in test vehicles
3. Experiment implementation
4. Weighting questionnaire result
5. Rating.
6. OWS value calculation
7. Workload assessment categorization
8. Analysis and discussion

3 Results and Analysis

Figure 1 illustrates the situation and the complexity of driving across the city road, rural road and highway in Malang, Indonesia. The figure exhibits hazards and constraints encountered by drivers passing through the city such as pedestrian crossing the street, parking vehicles, etc.. On the other hand, rural road tend to have simple road network condition, more drivers interact with other vehicles. As for highway riders concentrate more on their respective vehicles and road structure. It involves streets' ascent, descent, turn or bump.

Data on mental workload using NASA-TLX method used six indicators to determine mental workload experienced by the driver. Indicators are Mental Demand (MD), Physical Demand (PD), Temporal Demand (TD), Performance (P), Frustration level (F), Effort (EF). On the other hand, weighting stage presents 15 pairs of mental load indicator, to be filled by the respondent (driver) by circling one of the indicator pairs which they thought to be more dominant or important on driving activity. The weighting factor of mental workload constraints on city road examination result is exhibited in Table 1 with average MD = 3.62, PD = 1.31, TD = 1.81, P = 1.50, F = 3.92, and EF = 2.85. Based on questionnaire records on rural road, mental load factor value obtained were as follows: MD = 3.35, PD = 1.81, TD = 1.88, P = 1.73, F = 3.46, and EF = 2.77. Value obtained for highway section was: MD = 3.19, PD = 1.54, TD = 2.35, P = 2.88, F = 2.50, and EF = 2.77. Based on research data, driving in city road requires more and dominant mental effort compared to physical needs. It also has a high level of stress which is supported by the value of low job satisfaction in completing the task.

Figure 1. Description of the driving situation on three different road conditions: a. City Road, b. Rural Road, C. Highway.

Table 1. Weighted Data of NASA TLX Questionnaire on city road

Driver	Indicator						Total Score
	MD	PD	TD	P	F	EF	
Driver 1	4	2	0	1	5	3	15
Driver 2	3	1	4	0	5	2	15
-	-	-	-	-	-	-	-
Driver 25	3	2	2	2	3	3	15
Driver 26	4	1	1	2	3	4	15

Rating stage follows weighting stage. At this stage, a rating of 1-100 scale is given for each indicator according to the circumstances experienced by the driver. Rating distribution comparison result for all indicators on each track type is exhibited in Figure 2 below. The figure exhibited driver driving through city road has the greatest mental burden. They possess the highest frustration, effort, and mental demand. On the other hand, performance has the lowest value. Drivers passing through the rural roads tend to give a good performance, despite experiencing a high level of frustration and hard effort. The overall highway has a positive impact on drivers which exhibits low levels of frustration and physical demand but results in high satisfaction of driving achievement.

The following step is to calculate Overall Workload Score (OWS) which aims to obtain the value of mental workload on each indicator. The category of workload assessment consists of three levels: low with the interval scale of 0-9, medium for interval scale is 10 - 29, slightly high at interval scale of 30 - 49, high at interval scale 50 - 79 and very high with interval scale of 80 - 100. Below is an example of a calculation recapitulation of OWS driver who drives on city road.

$= ((80 \times 4) + (75 \times 2) + (65 \times 0) + (85 \times 1) + (85 \times 5) + (90 \times 3))/15$

$= (320 + 150 + 85 + 425 + 270) / 15$

DOI: 10.1051/matecconf/201713602007

= 1250 / 15
= 88.33

Based on calculation results above, driver mental workload in a subjective measurement can be classified in "very high" category. Overall it can be classified that the drivers traveling on city road and rural road will experience a mental burden of 3.85% in "very high" category and 96.15% in "high" category. Drivers traveling in highway are 100% categorized in "high" workload. Car technology, proper driving training and improving driver's own health level.

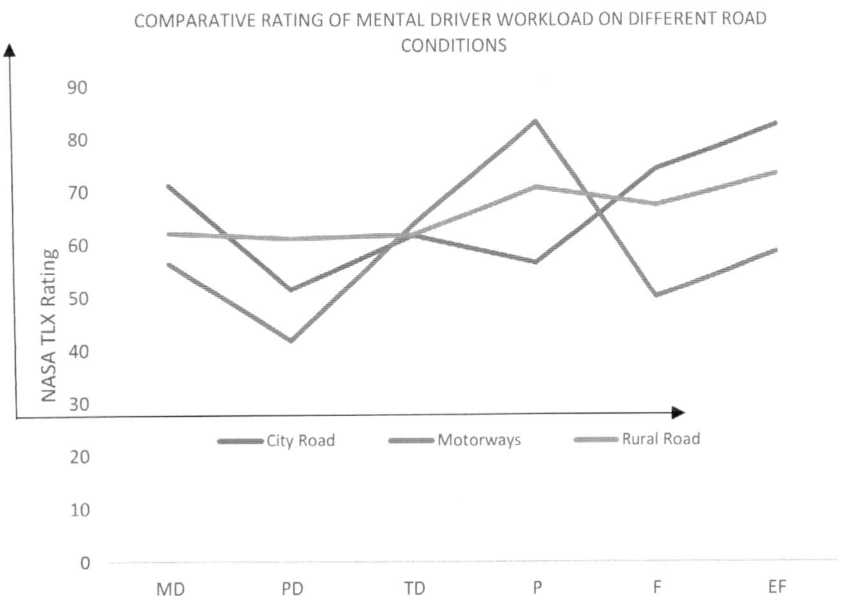

Figure 2. Rating comparison on each NASA TLX indicators in different road conditions

4 Conclusions

This research has successfully demonstrated how NASA TLX can be utilized to measure subjective mental workload on drivers crossing different types of streets (city road, rural road, highway) around Malang City, Indonesia. Based on measurement results, it was exhibited that almost every respondent have a mental workload in "high" category. Mental demand frustration factor becomes the main cause of high mental work load, then followed by performing hard effort to reach their purpose. Physical needs have not become a serious consideration compared to other factors to achieve driver satisfaction. Therefore special handling based on driver experience is required. It is recommended to improve infrastructure, transport regulations, and innovation of safe and comfortable driving technologies.

Acknowledgements

The researcher expresses their gratitude to Ministry of National Education of the Republic of Indonesia for supporting this paper. The authors are also grateful to the Work Design and Ergonomics Laboratory, Industrial Engineering Department, University of Brawijaya, Malang, Indonesia for their extraordinary support.

References

1. Fastenmeier, W. & Gstalter, H., SC, **45**, 952-979 (2007).
2. Fuller R., AAP, **37** 461–472 (2005).
3. Schlegel, R.E, *Automotive Ergonomics*, Taylor & Francis, London (1993).
4. Bhise, V. D, *Ergonomics in the Automotive Design Process*, Boca Raton, CRC Press, (2012).
5. Verwey, W. B. & ZaideL, D. M., AAP **31**(3), 199-211 (1999).
6. Shih-Nan Lu, et al., TJSE, **13** (4), pp. 395404 (2010).

7. Alex Cao et al., BRM, **41**(1), 113 -117 (2009).
8. Ambarwati L. et al., JRS, **4**(2), (2010).

Effect of Wave Propagation in a Distributed Energy Absorption System for Trains

Zhaoyang Ding[1] Zhijun Zheng[1,a] and Jilin Yu[1]

[1]*CAS Key Laboratory of Mechanical Behavior and Design of Materials, Department of Modern Mechanics, University of Science and Technology of China, Hefei 230026, China*

Abstract. The influence of elastic wave propagation along the carriage in high-speed trains on the coordination of energy absorber components is analyzed in this paper. A simplified model is developed to simulate a "like to like" impact scenario of trains with distributed energy absorbing system. A rod made of elastic material is used to simulate a carriage, while the rigid-perfectly plastic-locking (R-PP-L) model is employed to describe the material of the energy absorber. The problem is simplified to a one-dimensional case. The response of absorption components during collision is divided into several stages and the corresponding governing equations of the impact response are obtained and numerically solved. A typical platform phenomenon of response in all stages is observed. The results of theoretical analysis are compared with those of finite element simulation and good agreement is achieved.

1 Introduction

The active and passive safety technology has received much attention with the development of high-speed trains. The passive safety technology is an efficient means of reducing losses in accidents when the active safety technology is failure. Several collision experiments of actual sized trains were implemented to examine the crashworthiness of train [1]. Recently, numerical simulation combined with theoretical analysis becomes the most commonly used and economical measure. The optimal methods for crashworthiness design [2, 3] and the safety of passengers [4, 5] have received much research interest. Multi-body dynamics analysis was widely used for train crash analysis [6, 7].

A train is a so long structure system that the length involved in the event increases with the passage of time and the impact load propagates from the front to the rear of a carriage in the form of elastic wave. Therefore, in order to develop the energy absorbing system, the effect of stress wave propagation along the train should be considered.

In this study, a simplified one-dimensional model, which takes the distribution of energy absorbing devices and the elasticity of the train carriages into consideration, is developed for a "like to like" impact scenario of trains. The governing equations of the impact response are obtained and numerically solved. The results are then compared with those of finite element (FE) simulation.

2 A simplified model

During collision, the axial action plays a decisive role both in load transfer and in compression failure behavior of energy absorbing components. Therefore, the train can be simplified to a one-dimensional model and can be investigated based on a one-dimensional stress wave theory, when neglecting the various influences of the detailed structures on the load transmission and energy absorption. Elastic rods are employed to represent the carriages, connected by plastic layers, which represent energy absorbing devices.

[a] Corresponding author: zjzheng@ustc.edu.cn

As the plastic effect is more concerned rather than the elastic effect of energy absorbing device and the time scale of elastic wave propagation, the rigid-perfectly plastic-locking (R-PP-L) model [8] is employed to characterize the material response of energy absorbing device, as shown in Fig. 1, with the plateau stress and the densification strain corresponding to the mean crash force and the stroke efficiency, respectively. On the other hand, linear-elastic material is employed for carriage. As shown in Fig. 2, the simplified model of each carriage with energy absorbers is made up of an elastic compartment, and an energy absorption layer at the back-end, which is named as a distal layer. For simplicity, we put the energy absorbers between the nearby carriages together into the back end. For the head car, there is an extra energy absorbing layer at the front-end, which is named as a proximal layer.

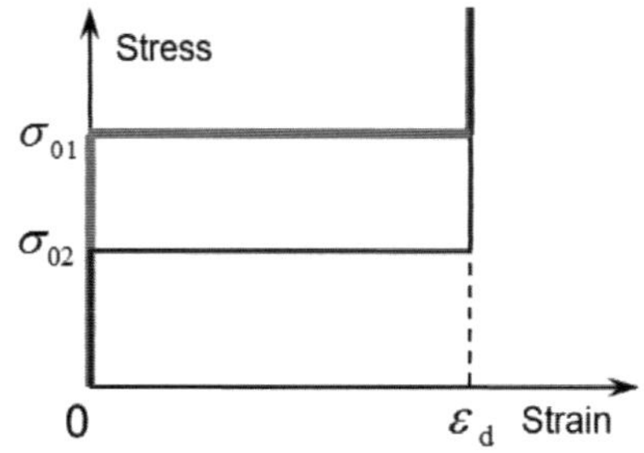

Figure. 1 The R-PP-L model.

To be noted, the energy-absorbing components in a distributed energy-absorbing system need to meet a certain distribution rule that the subsequent energy absorbing components should start to work before the one at the impact end is fully compacted. To meet this requirement, the plateau stress of the proximal layer (σ_{01}) should be higher than that of the distal layer (σ_{02}), as illustrated in Fig. 1.

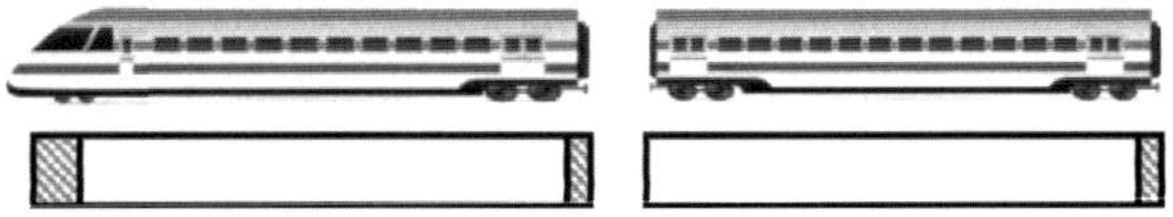

Figure. 2 A schematic diagram of the simplified model.

A collision scenario is analyzed in this study: a simplified model containing a head car and a carriage with an initial velocity impacting on a fixed rigid wall. This scenario simulates a "like to like" impact, in which the two trains with the same initial velocity and mass strike each other, as shown in Fig. 3.

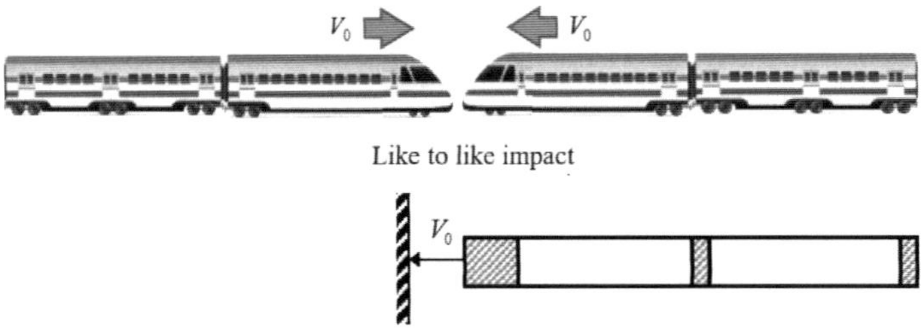

Figure. 3 The collision scenario of actual train and the corresponding simplified model.

3 Theoretical derivation

In this section, the collision scenario is investigated in detail and the governing equations of motion are analyzed based on Lagrangian coordinates. A schematic diagram of simplified model and the deformation process is presented in Fig. 4.

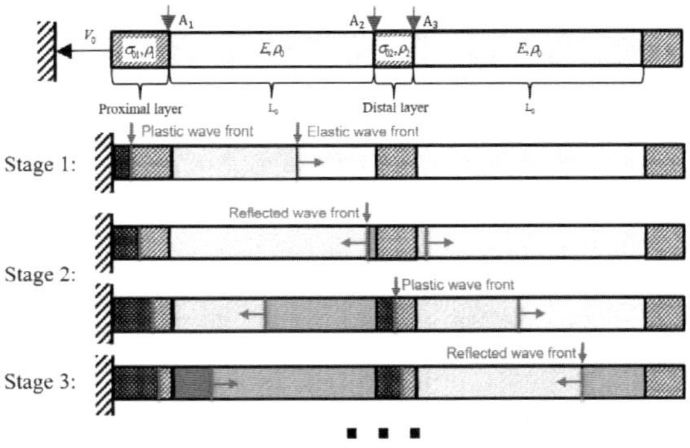

Figure. 4 A schematic diagram of the deformation process.

The parameters of the energy absorption layers used in the analysis of the simplified model are: the plateau stress σ_{01} and σ_{02} are 150 MPa and 100 MPa, respectively; the density and length of the proximal layer are $\rho_1 = 600$ kg/m^3 and $L_1 = 1$ m, and those of the distal layer are $\rho_2 = 400$ kg/m^3 and $L_2 = 0.5$ m; the compaction strain is $\varepsilon_d = 0.8$ for both layers. The length of the elastic rod is $L_0 = 25$ m, which has a Young's modulus $E = 66000$ MPa and a density $\rho_0 = 2700$ kg/m^3. The initial velocity of the simplified model of train is $V_0 = -20$ m/s.

Parameters to be solved are: the compaction length of the proximal layer Φ_1 and that of the distal layer Φ_2, the particle velocity v_{R1} at interface A_1 associated to the right traveling wave from interface A_1, particle velocity v_{L1} at interface A_2 associated to the left traveling wave from interface A_2, and particle velocity v_{R2} at interface A_3 associated to the right traveling wave from interface A_3. Other unknowns including the stress and particle velocity distribution can be obtained from these parameters.

Elastic wave propagation is the key for the load and impact energy transmissions between the proximal layer and the distal layer. So the analysis process can be divided into several stages by the characteristic time t_n (n = 1, 2, ...) when elastic waves reach corresponding interfaces

$$t_n = nL_0 / C_0 , \quad (1)$$

where C0 is the elastic wave speed, given by

$$C_0 = \sqrt{E / \rho_0} . \quad (2)$$

Stage 1 (0 < t < t_1)

When the front end hits the rigid wall, the proximal layer begins to collapse and a compaction wave starts propagating in the proximal layer. The stress ahead of the compaction wave front becomes σ_{01} which pushes the rigid part of the proximal layer and decelerates it immediately. The stress on the interface A_1 rises and approaches σ_{01}. In the meanwhile, an elastic wave initiates and travels from interface A_1 to interface A_2 through the elastic rod at speed C_0.

For interface A1, the stress σ_{A1} can be expressed by the particle velocity v_{R1}. Using the compatibility relation of elastic wave and the initial conditions $v_{R1}(0) = 0$ and $\sigma_{A1} = 0$, we have

$$\sigma_{A1} = Ev_{R1} / C_0 \quad (3)$$

By using the conservation of the mass of the proximal layer, the momentum theorem of the undeformed region of proximal layer and Eq. (3), the governing equations are obtained as

$$\frac{d\Phi_1^{(1)}}{dt} = \frac{0-(V_0 + v_{R1}^{(1)})}{\varepsilon_d} \quad (4)$$

and

$$\frac{dv_{R1}^{(1)}}{dt} = \frac{\sigma_{01} - Ev_{R1}^{(1)} / C_0}{\rho_1 \left(L_1 - \Phi_1^{(1)}\right)} \quad (5)$$

with the initial conditions $\Phi_1^{(1)}(0) = 0$ and $v_{R1}^{(1)}(0) = 0$, where superscript (n) denotes the n-th stage of the response (n = 1, 2, ...).

Stage 2 ($t_1 \le t < t_2$)

When the elastic wave reaches interface A_2, an elastic precursor wave travels through the distal layer with an infinite speed. Interface A_3 starts to decelerate and an elastic wave propagates right. The response of interface A_2 due to the right traveling wave is corresponding to that of interface A_1 with only a time delay of

$$\Delta t = L_0 / C_0 . \tag{6}$$

The velocity and stress of interface A_2 can be expressed as

$$v_{A2}^{(2)}(t) = V_0 + v_{R1}^{(1)}(t - \Delta t) + v_{L1}^{(2)}(t) \tag{7}$$

and

$$\sigma_{A2}^{(2)}(t) = \sigma_{R1}^{(1)}(t - \Delta t) + \sigma_{L1}^{(2)}(t) . \tag{8}$$

Originally, σ_{A2} is lower than σ_{02} and the distal layer remains undeformed. The equations of motion is given by the momentum theorem of distal layer

$$\frac{dv_{L1}^{(2)}}{dt} = -\frac{2Ev_{L1}^{(2)}(t)}{C_0\rho_2 L_1} - \frac{dv_{R1}^{(1)}(t-\Delta t)}{dt} \tag{9}$$

with the initial condition $v_{L1}(t_1) = 0$.

The time when σ_{A2} reaches σ_{02}, denoted as t_c, can be determined by Eqs. (8) and (9) and dealing with σ_{L1} similarly as Eq. (3). Afterwards the distal layer begins to be crushed. The compaction stress on the collapsing wave front can be obtained by the momentum theorem of the infinitesimal there and the mass conservation in distal layer

$$\sigma_2(t) = \sigma_{02} + \rho_2\left(v_{L1}^{(2)}(t) + v_{R1}^{(1)}(t-\Delta t) - v_{R2}^{(2)}(t)\right)^2 / \varepsilon_d . \tag{10}$$

The equations of motion are obtained by the mass conservation, the momentum theorem of the deformed and undeformed regions of the distal layer

$$\frac{d\Phi_2^{(2)}}{dt} = \frac{v_{L1}^{(2)}(t) + v_{R1}^{(1)}(t-\Delta t) - v_{R2}^{(2)}(t)}{\varepsilon_d} \tag{11}$$

$$\begin{aligned}\frac{dv_{L1}^{(2)}}{dt} = &-\frac{dv_{R1}^{(1)}(t-\Delta t)}{dt} + \frac{1}{\rho_2\Phi_2^{(2)}(t)}\Big[E\left(v_{R1}^{(1)}(t-\Delta t) - v_{L1}^{(2)}(t)\right)/C_0 \\ &-\sigma_{02} - \rho_2\left(v_{L1}^{(2)}(t) + v_{R1}^{(1)}(t-\Delta t) - v_{R2}^{(2)}(t)\right)^2/\varepsilon_d\Big]\end{aligned} \tag{12}$$

$$\frac{dv_{R2}^{(2)}}{dt} = \frac{\sigma_{02} - Ev_{R2}^{(2)}/C_0}{\rho_2\left(L_1 - \Phi_2^{(2)}\right)} \tag{13}$$

with the initial conditions $\Phi_2^{(2)}(t_c) = 0$, $v_{L1}^{(2)}(t_c) = v_{R1}^{(1)}(t_c - \Delta t) - C_0\sigma_{02}/E$ and $v_{R2}^{(2)}(t_c) = 2v_{R1}^{(1)}(t_c - \Delta t) - C_0\sigma_{02}/E$, where Eq. (10) is used.

Stage 3 ($t_2 < t < t_3$)

The left-traveling wave reaches interface A_1 at t_2, then the velocity response of interface A_1 depends on the superposition of the left-traveling wave and a new right-traveling wave, which cause the equations of motion to be different from those of Stage 1. We have

$$\frac{d\Phi_1^{(3)}}{dt} = \frac{0 - \left[V_0 + \left(v_{R1}^{(3)}(t) + v_{L1}^{(2)}(t-\Delta t)\right)\right]}{\varepsilon_d} \tag{14}$$

and

DOI: 10.1051/matecconf/201713602008

$$\frac{\mathrm{d}v_{\mathrm{R1}}^{(3)}}{\mathrm{d}t}=\frac{\sigma_{01}-E\left(v_{\mathrm{R1}}^{(3)}(t)-v_{\mathrm{L1}}^{(2)}(t-\Delta t)\right)/C_0}{\rho_1\left(L_1-\Phi_1^{(3)}(t)\right)}-\frac{\mathrm{d}v_{\mathrm{L1}}^{(2)}(t-\Delta t)}{\mathrm{d}t} \tag{15}$$

with the initial conditions $\Phi_1^{(3)}(t_2)=\Phi_1^{(1)}(t_2)$ and $v_{\mathrm{R1}}^{(3)}(t_2)=v_{\mathrm{R1}}^{(1)}(t_2)+v_{\mathrm{L1}}^{(2)}(t_2-\Delta t)$.

For simplicity, only the theoretical derivations of the first three stages are demonstrated here. The derivations for more stages follow the similar rules and can be obtained.

4 Results and discussion

4.1 Numerical solution of the theoretical model

The fourth-order Runge-Kutta scheme is employed to solve the governing equations with the initial conditions given above. Typical velocity-time history and stress-time history at the three interfaces are shown in Fig. 5. As can be seen from this figure, the speed of the three interfaces can be described by several stages and keep approximately constant in each stage. For the definition of stage employed in this paper, interfacial velocities change once every two stages, which is equal to a time length of an elastic wave traveling back and forth between the two interfaces. The stress of the interfaces is almost equal to the corresponding plateau stress during collision and drops down afterwards.

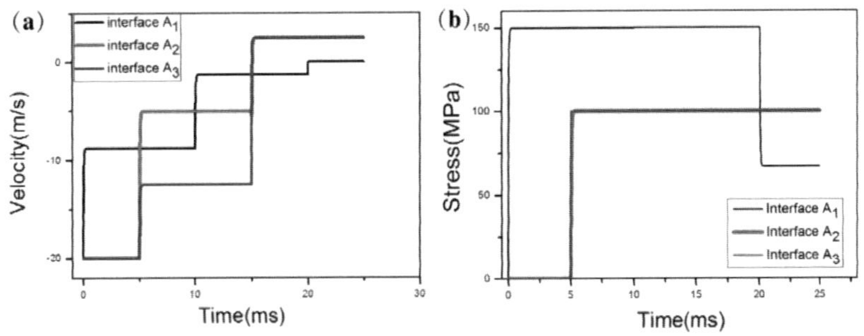

Figure. 5 Typical velocity-time history (a) and stress-time history (b) of interfaces A_1, A_2 and A_3.

4.2 Comparison with finite element results

Finite element method with ABAQUS/Explicit code is employed to verify the correctness of the theoretical analysis. The FE model is set up with the same parameters as the simplified model. Both the energy absorbing layers and the elastic connecting rods are modeled with C3D8 solid elements, of which the average length is set to be about 1mm through a mesh sensitivity analysis. Crushable-foam-material and elastic-material are employed for the two layers and the elastic rods, respectively.

Local response of velocity and stress in the elastic rod is resulted from the superposition of two elastic waves, which can in turn verify the status of the two interfaces in each stage. Time histories of stress and velocity for selected Lagrangian locations in the first elastic rod are presented in Fig. 6. The thick black line is the result of the theoretical model for the midpoint. The thin and color lines are FE results for four selected positions, whose Lagrangian coordinates are 0, $L_0/4$, $3L_0/4$ and L_0.

 DOI: 10.1051/matecconf/201713602008

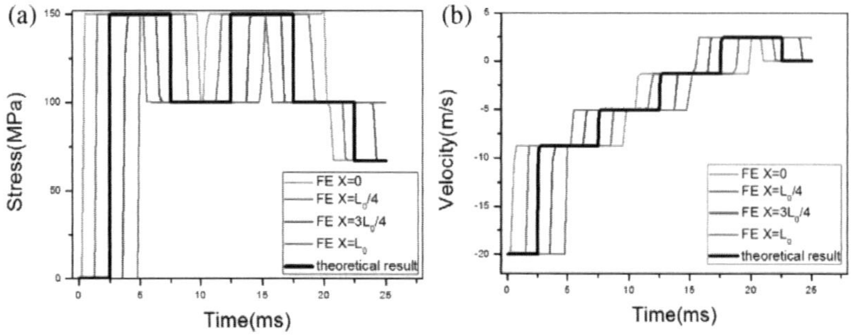

Figure. 6 Time histories of stress (a) and velocity (b) for selected Lagrange locations in the elastic rod.

Both the velocity and stress of the three interfaces are presented in Fig. 7. The theoretical and FE results are distinguished by lines without and with symbols, respectively.

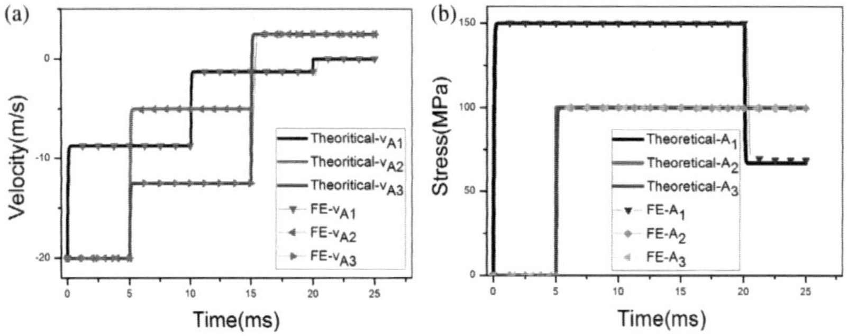

Figure. 7 Comparison between the results from theoretical analysis and FE simulations.

The observed agreement between the theoretical predictions and the FE results presented in Fig. 6 and Fig. 7 verifies the proposed theoretical approach. Minor difference of response time is result from that an infinite velocity of elastic wave front in two layers cannot be realized in the finite element simulation.

5 Conclusions

In this paper, a simplified model considering elastic wave effect is introduced to analyze the performance of the distributed energy absorbing system of trains. The governing equations for different stages of the response of the whole system in the course of collision are obtained and numerically solved. Typically segmented and platform typed phenomena of response are observed. The theoretical predictions are compared well with the finite element simulation results.

The proposed model is expected to be used for the analysis of the coordination of energy absorber components and preliminary design of the distributed energy absorption system for trains.

Acknowledge

The research reported herein is supported by the National Natural Science Foundation of China (Project No. 11372307), which is gratefully acknowledged.

References

1. P. Drazetic, R. Tassin, Y. Ravalard, et al, P I MECH ENG F-J RAI, **209**, 11-17 (1995)
2. J. P. Dias, M. S. Pereira, COMPUT STRUCT, **82**, 1371-1380 (2004)
3. G. J. Gao, H. Q. Tian, Int. J. Crashworthiness, **12**, 21-28 (2007)

4. J. W. Simons, S. W. Kirkpatrick, Int. J. Crashworthiness, **4**, 121-132 (1998)
5. S. W. Kirkpatrick, M. Schroeder, J. W. Simons, Int. J. Crashworthiness, **6,** 95–106 (2001)
6. J. A. C. Ambrósio， M. F. O. S. Pereira， J. P. Dias, NONLINEAR DYNAM, **10,** 359-379 (1996)
7. H. S. Han, J. S. Koo, VEHICLE SYST DYN, **40,** 435-450 (2003)
8. S. R. Reid, C. Peng, INT J IMPACT ENG, **19**, 531-570 (1997)

Design Strategy of Information Construction Based on User Participation

Jundong Gao[1,a]

1Department of industrial design, Huazhong University of Science and Technology

Abstract. In the design of the Internet products, the information construction is one of the important factors to determine that whether a product is friendly. It is a challenge for every information architect to discover new innovations on the basis of established user habits. In practice, we often found that the development of many new functions did not meet the needs of users and many functions would be improved by users themselves, the fact of which shows the strong desire of users to participate in the development of products. Therefore, this article proposes an information architecture method based on the "users' partially involved", that is, when the main framework of a product is determined and the detailed architecture of part of the functions is proposed by users themselves, builders will help users to dig and organize their requirements and eventually integrate them into the design of architecture. In order to verify the effectiveness of this method, the article takes the virtual products for social service based on "surrounding geographic information" as an example. A comparison experiment was conducted on the effectiveness of the design of user-participant information architecture, and the experimental results were evaluated.

1 Literature Review

The architecture of information is a task that designers, developers, and content decision makers often need to face. As the bones of product, we should try to keep it simple and clear, rather than being easily changed, making users overwhelming[1]. Therefore, the information architecture of the product, which is of great significance to the product itself and the users' experience, needs a scientific and logical way to combine.

Scholars began to research the methodology of information architecture in 1970s, long before the emergence of the website and smart phone App. It originated from the methodology which is still used by many UX designers nowadays, and has three main categories, Library Information Science, Cognitive Psychology and Architecture[2]. The Bibliography and Archives of Library Information Science provide the organization method for information architecture.

In the research of design category of information architecture, An organization called UXbooth divides the content of information architecture into seven categories: User Research and Analysis, Create Navigation and Hierarchy, Framework, Tag, Taxonomy and Metadata, Data Modeling Parallel Literature[3], In the previous architecture design, information architects often construct the content architecture diagram and the wireframes with the procedure according to the data analysis, and classify the function level of the sub information in the wireframe, share the data with team in the form of prototype, pictures or suggestions, so that we can show the typical users of products, the logic and interaction path of functional requirements, and the way to use App.

But in this architecture approach, the degree of user involvement in product design is shallow, the form tends to be passive, Users mainly accept interviews, accept observation, fill in questionnaire and do usability test of the product according to the task set by the researcher, most of them make passive choices from different schemes, and express their views. Whether it is in the user's subjective feelings, or in the designers' and researchers' cognition, the users are objects of study, are cooperative party[4]. In

[a] Corresponding author: gaojunku2008@163.com

recent years, with the form of users involved in the design of is used in the new design strategy, the roles of designer and user have subtly changed in comparison with tradition: the user becomes the product designer and changer, while designers and development engineers are more like coordinators and observers in PD session, perceptually getting the users' first-hand information. In this process, Researchers can tap users' consciousness and demand from a deeper perspective as the main organizers of the PD. Compared with the traditional design pattern, PD extremely emphasize taking the initiative and enthusiasm of the user who is no longer just passive from different schemes to make choice, to express their views, but really participates in the prototype design, designs in the way of brainstorm, and even be absorbed into the design team, working with designers in a short period of time. With the help of the materials provided by the experimenter, the user can provide the design proposal and explain his creative ideas. In the whole process, the user feels that he is working with the designer to create and solve the problem, he is the changer of the product.

This paper will take the mobile application of smart phone as the research object, to explore the design patterns in which the users will also partly participate, present a method of information architecture of mobile product, and verify the effectiveness of this method through the A/B test.

2 Pre-study - research purposes and research object analysis

The aim of this study is to test the effectiveness of information architecture design method with partial participation of user, Information architecture design method in the past, the APP information architecture is often through direct extraction of competing goods architecture, and then combined with the function design of the product itself, this method is based on the Angle of product architecture to think. And the proposed approach is obtained by using the method of user part of information architecture, he is a need to combine product and user mental models it is concluded that the method of architecture itself. The study object is the APP based on geographic location service, and the main function of such APP is to help people making online food ordering and seat reservation and arrange offline dining accordingly, to save user's time of waiting for empty seat and queuing, so that help people solve their dinning demand in restaurant through queuing. on one hand, it was such a behavior developed from actual living behavior to the internet then to mobile internet, with the typical significance; on the other hand, the reason is that, for such behavior of online food booking and offline dining, the user may have quite large difference in the original intention of product's developers, for product development, the developers may consider more about product itself, but the user may consider that the software usage scenario and using habit shall have the same important significance with the experience of product itself.

3 Experiment I - the information architecture design based on competing product thought

The aim of this study is to use general method to abstract the information architecture of product A, the abstract method is usually the methods provided by product's decision-maker, without the protruded content irrelevant to core function, it can enable the user completing the task and fulfill the demand of product itself accordingly.

First of all, we selected to sort the logic of original product, made information architecture analysis and task process analysis, to perform segmentation firstly according to the application's functional module, to abstract the architecture of dinning part and refine it in single function, further to draw all the functions in the tree structure. From such a task process we can find out user's core using function, then to finally result in the architecture based on product thinking by combining the core functions and according to the structure of competing product, and manufacture the prototype A accordingly.

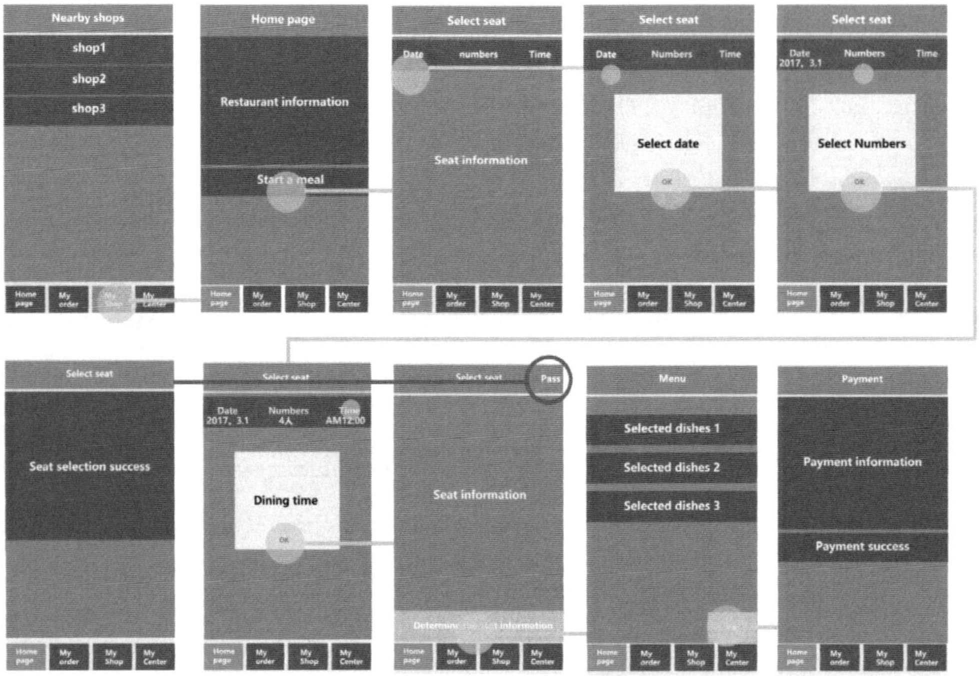

Figure 1. Test prototype A

4 Experiment II - the information architecture strategy design based on partial participation of users.

This study is the focus of the whole experiment, in order to introduce the information architecture approach for the part of the user.

The first step is to find three products based on geographic location as the competing products to analyze the object of analysis. They are respectively the three mainstream geographic information service applications as Meituan.com, Alipay's public praise and Dianping.com. Unlike previous methods, the aim of our competing product analysis is to analyze the generality and difference among the architectures, in other words, it refers to make better innovation with the premise that do not impede user's habit. On one hand, to analyze the already formed habits of the user in the part of generality, for example, if the navigation in certain level and logic of core function in many competing products both used the same or similar processing method, this means that such part of architecture or user's using habit was already stable, without the enough assurance, it cannot make the blind innovation for representing the difference, otherwise it may bring the huge study cost for the user and influence the usage experience of the product. On the other hand, to find out the innovative point of the architecture in the differences.

Through the analysis on the competing products, we have found the generality of the three architectures: the first one, The user's mental model and usage logic have been relatively stable in the organization architecture of selecting location information and time information, and in the user core usage process, users' mental model and using logic was relatively stable, the user has also formed the solid habit in the two using methods of seat selection and food booking, during the architecture sorting these cannot be changed easily. The second, the overall layout of the information architecture in the three competing products are basically the same, especially the core usage process based on food booking, for the content layout in home page, they both used the label navigation, so we've also used the same organization method to control experiment effect. The third, the organizational architecture of competing products did control the vertical level depth in 4 stories, but had quite little differences in horizontal range, "Alipay" and Ele.me maintained the four dimensions range, but "Meituan" gave up the "discover" bar, maintained in three dimension range. From the perspective of cognitive psychology, we'd control user's vertical and horizontal cognitive load within 4 stories approximately.

DOI: 10.1051/matecconf/201713602009

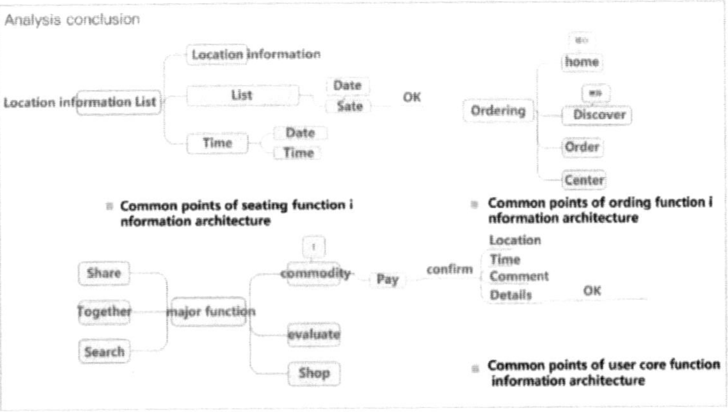

Figure 2. Similarities and differences in architecture

The second step, we've used card classification method to collect the information architecture based on user mental model.

First of all, we handled the target users with disorganized end information through architecture A, the designer's task was to interpret the function of each card, ensuring that the users have had completely understand the function of these cards, then the participants were organized and classified by themselves, and also arrange the name for each category. The process of guiding user to classify the information into different grades may need 2 or 3 times, so as to ensure the user classification being not too concentrated or dispersed. Then the relative functional cards were classified according to their importance degree, during this process the record was also maintained very well, finally it can result in the classification based on user mental model, by comparing the difference in the competing products and user mental model classification to record them, finally to organize the organizational structure based on user mental model. At last, to disorganize the functional elements of relative web pages at the place with such difference, let the users make the combination within the range provided by us, the aim is to sort out the related function of important level and find out the problems and needs, We have come to some conclusion in the experiment that users often classify order and reservations as one category, After an interview analysis, we decided to classify the booking and ordering as a class and maintain them at the same level .In the primary navigation, Product decision makers hope keeping "my shop" to facilitate the promotion of products, But this classification makes the user confused in the process of using the product, After the interview found that the user needs is to choose from a number of stores to enter, so the choice of business and determine the target business is a complete logic, you should not split their classification, so in the back of the structure we simplify one Level navigation for three categories and finally result in the architecture with partial participation of users, and make the prototype B accordingly.

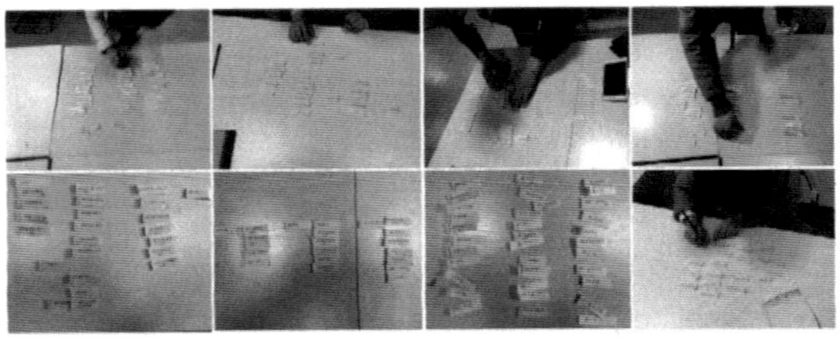

Figure 3. User participation in design process

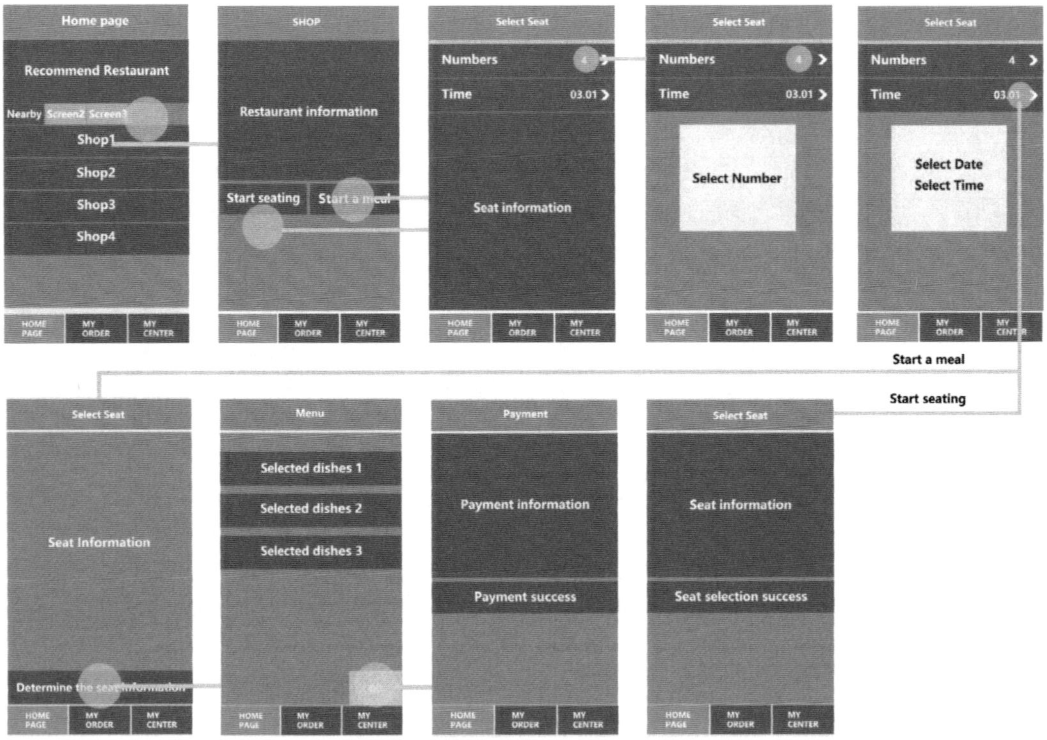

Figure 4. Test prototype B

5 Usability testing and result analysis

The main aim of this study is to discuss the effectiveness of design strategy with partial participation of users, the study methods include the objective performance of prototype A and B (product usability test) and subjective feeling (USE questionnaire)

The independent variable in this experiment can be divided in four task categories, they are respectively the completion of seat booking, completion of food booking, selecting restaurant nearby, find the completed order, the control variables are the equipment platform, visual factor. Based on the independent and control variables, we've used the product usability test and USE questionnaire to collect the performance measurement of participants completing the set task, the main dependent variables include the task successful degree, completion time of the task, completion efficiency, error rate and effectiveness, usability, learnability and satisfactory degree.

The participants in this experiment were the 15 college students whose age varied between 18 to 25, it included 8 females and 7 males, 9 of them were studied in the major of industrial design, other 6 came from the majors of mechanics, materials respectively. All the participants are the user who often used APP in daily life.

From the overall level it can be seen that: 1) there's difference in user experience comprehensive assessment of prototype A and B in 4 tasks, prototype B's comprehensive score was higher than that of A, the reason may be that the prototype B based on user mental model can even better match the using habits of the users. 2) in task 1, prototype A's effectiveness and usability were significantly better than prototype A, the reason may be that in prototype A the information level of the function "seat reservation" was intentionally weakened in product's consideration dimension, the user may pay more attention on food booking but ignore the functional position of seat reservation. 3) in task 2, the comprehensive score of the two prototypes are almost the same, the reason may be that it both considered user's core usage process in prototype A and B.

	learning-ease	using-ease	Satisfaction	Effectiveness	Overall merit
1	4. 39	3. 53	4. 47	3. 91	4. 075
2	3. 65	4. 11	3. 63	4. 62	4. 0025
3	4. 15	3. 72	3. 3	3. 55	3. 68
4	3. 38	4	4. 12	4. 29	3. 9475
5	3. 62	3. 38	3. 79	4. 02	3. 7025
6	4. 03	4. 19	4. 2	4. 39	4. 2025
Average	3. 87	3. 82166667	3. 918333333	4. 13	3. 935
Median	3. 84	3. 86	3. 955	4. 155	3. 975
Geometric me	3. 854392756	3. 80970806	3. 898692364	4. 114961964	3. 930416884
standard dev	0. 380893686	0. 32896302	0. 425742489	0. 382151802	0. 2073403

Table 1. Test results of prototype A

	learning-ease	using-ease	Satisfaction	Effectiveness	Overall merit
1	4. 56	4. 79	4. 18	4. 32	4. 4625
2	4. 09	4. 3	4. 68	3. 6	4. 1675
3	4. 21	3. 89	4. 11	3. 93	4. 035
4	4. 18	4. 53	3. 72	4. 68	4. 2775
5	3. 95	4. 38	4. 05	4. 32	4. 175
6	4. 14	4. 1	3. 6	4. 27	4. 0275
Average	4. 188333333	4. 33166667	4. 056666667	4. 186666667	4. 190833333
Median	4. 16	4. 34	4. 08	4. 295	4. 17125
Geometric me	4. 184311367	4. 32200193	4. 041979161	4. 172433069	4. 188224027
standard dev	0. 203707306	0. 31669649	0. 381715426	0. 373077293	0. 163138489

Table 2. Test results of prototype B

6 References

1. Norman, D A, The Design of Everyday Things[M]. New York: Doubleday, 1990
2. Lindgard G, Usability Testing and System Evaluation: A Guide for Designing Useful Computer Systems, Chapman and Hall, London, U.K, 1994
3. Tian Lei, Shuaili Wei. An Exploration of Relationships between Culture Images and User Experience of Gesture Interaction. In: Lasse Berntzen, Vestfold University College eds. CENTRIC 2013. Venice. 2013.pp 112-117.
4. Liu, Y.C., Ma, Y.H.: The Behavior Analysis of Product Negative Word-of-Mouth Spread on Sina Weibo. Human Centered Computing, pp. 284-295(2016)

Optimization of Mang[a]la Hydropower Station, Pakistan, using Optimization Techniques

Muhammad Zaman[1, 2,a,]Shouqi Yuan[1], Junping Liu [1], Ijaz Ahmad[3] ,Noman Ali Buttar[4], Muhammad Saifullah[5] and Muhammad Imran. Nawaz[6]
[1] *Research center of Fluid Machinery Engineering & Technology, Jiangsu University, P.R. China.*
[2] *Department of Irrigation and Drainage, University of Agriculture, Faisalabad, Pakistan.*
[3] *Center of Excellence in Water Resources, University of Engineering & Technology, Lahore, Pakistan.*
[4] *School of Agricultural Equipment Engineering, Jiangsu University, P.R. China.*
[5] *Department of Structures & Environmental Engineering, University of Agriculture, Faisalabad, Pakistan.*
[6] *School of Environment and safety Engineering, Jiangsu University, P.R. China.*

Abstract. Hydropower generation is one of the key element in the economy of a country. The present study focusses on the optimal electricity generation from the Mangla reservoir in Pakistan. A mathematical model has been developed for the Mangla hydropower station and particle swarm and genetic algorithm optimization techniques were applied at this model for optimal electricity generation. Results revealed that electricity production increases with the application of optimization techniques at the proposed mathematical model. Genetic Algorithm can produce maximum electricity than Particle swarm optimization but the time of execution of particle swarm optimization is much lesser than the Genetic algorithm. Mangla hydropower station can produce up to $59*10^9$ kWh electricity by using the flows optimally than $47*10^8$ kWh production from traditional methods.

Keywords: Mangla Reservoir, Hydropower, Optimization techniques, PSO.

1 Introduction

Optimal management and utilization of water resources is a serious and main issue in water scarce countries, which is also mainly affected by changing climate. Optimal allocation of water resources for adaptable use is also critical and complicated subject for water managers, because of inter-reliant and non-linear relationships. Consequently, optimal allocation of water resources, which is usually mathematical based, is a foremost area of study now a days in water resources advancement [1]. Computer based complex processes were established in 1970s to deal with such kind of problems [2,3]. Researchers exploited a lot of versatile techniques, linear to dynamical programming, to solve the water allocation and reservoir operation problems in the past [4,5]. Abundant water resources studies have made the use of these techniques for various problems [6,7]. Though, the nonlinear approaches require a lot of information about constraints and objective function, which make them inefficient.

Genetic algorithms (GA) are quest methods relay on natural selection mechanism that combine the fittest artificial survival with genetic operators. A randomly generated chromosome population, which represents the initial solution of the possible problem, goes through the GA. GA is an optimal search strategy, of a set of likely possible designs rather than a single design; that permits the Genetic algorithm to keep a multipoint observation on several locations within the solution space simultaneously, resulting in a high possibility of discovering the global optimum [8].

[a] Corresponding author : muhammad.zaman@uaf.edu.pk

 DOI: 10.1051/matecconf/201713602010

Numerous studies have utilized this technique to solve the water resources problems and found that GA is a promising technique [1,9,10]. Oliveira and Loucks (1997 a,b) used genetic algorithm to figure out the multi-reservoir operations [11], whereas Wardlaw,[12,13] and Sharif (1999;2000) also used GA in multi reservoir operation and concluded that GA is a robust technique and replacement of conventional optimization techniques. Van Zyl et al (2004) combines the GA with a hill climber search method and used it for optimal water distribution [14]. Wang et al (2011) utilized the GA for water resources problems and found it a comprehensive, versatile and best technique to find the optimal solution [15].

Another significant optimization technique named as Particle Swarm optimization (PSO), developed by Kennedy and Eberhart (1995), is becoming popular to solve optimization problems [16]. Numerous studies utilized this technique to solve water related problems and found it robust and best for getting best solutions in water resources [17].

Pakistan is facing water shortage issues from last decade and Mangla reservoir is one of the major power generation source in this electricity scarce country. Optimal electricity generation and utilization of available water in this reservoir is a complicated and critical issue now a day. To the best of authors knowledge, recently no work has been done in this area to get the optimal hydropower generation from the hydropower stations. There is a dire need to utilize available water resources and to get optimal electricity generation for the electricity short country. The presented study is a try to get optimal hydropower generation from Mangla hydropower station by applying two numerous GA and PSO algorithms at newly developed mathematical model for Mangla hydropower station.

2 Study area

Mangla hydropower is installed at Mangla reservoir situated at Jehlum river, which is a major tributary of Indus river. The Jehlum River receives water from major four of its tributaries named as Neelum, Poonch, Naran and Kanshi Rivers and dispose into the Indus River. The Mangla reservoir and hydropower station is the only one existing station, currently at the Jehlum River and meeting point of aforesaid tributaries. The hydropower is installed at the Mangla Dam, which is 147m high and about 3000m in length with a water storage capacity of about 9.12km^3. The installed hydropower station has a capacity of 1000MW with maximum efficient conditions. Mangla reservoir and hydropower station are shown in Figure 1 as below.

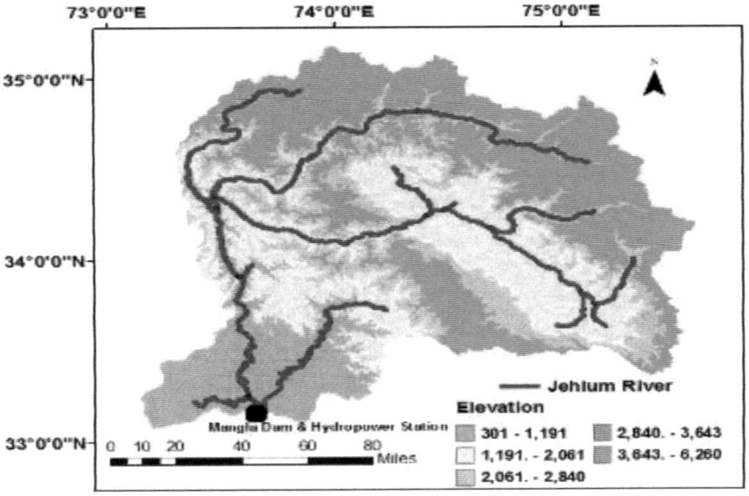

Figure 1: Location of the Hydropower station

3 Methodology

3.1 Mathematical model

A mathematical model has been developed and utilized for Mangla Hydropower station. The model consists of mainly constraints and objective function. Reservoir water level is taken as constraints and maximization of the hydropower as an objective function by the model. The objective function of the model and constraints are given below

$$E_{max} = \sum_{k=1}^{T} \sum_{k=1}^{M} A_k q_{kt} H_{kt} \Delta t \quad (1)$$

Whereas the constraints are given as below
Water balance equation

$$V_{k,t+1} = V_{kt} + (Q_{\lambda KT} - Q_{kt}) * \Delta t \quad (2)$$

Reservoirs discharge limits

$$Q_{kt,min} \leq Q_{kt} \leq Q_{kt,max} \quad (3)$$

Reservoirs storage volume limits

$$V_{kt,min} \leq V_{kt} \leq V_{kt,max} \quad (4)$$

$$N_{kt,min} \leq N_{kt} \leq N_{kt,max} \quad (5)$$

Where T is total time period within a year, T=12 ,M is the number of total reservoirs, A_k Coefficient for Power generation , E is maximum hydropower generation , $Q_{\lambda KT}$ is Inflow of reservoir k at time period t, m^3/s, H_{Kt} k reservoir average head at time period t, m, $V_{K,t+1}$ reservoir k volume at the end of time period t, $Q_{Kt,min}$ / $Q_{Kt,max}$ are minimum/maximum k reservoir outflow at time period t, m^3/s , $V_{It,min}$ / $V_{It,max}$ minimum/maximum k reservoir volume at time t, $N_{It,min}$ / $N_{It,max}$ are Minimum/maximum k hydropower installed capacity at time period t

3.2 Genetic algorithm

GA is an optimal search strategy, of a set of likely possible designs rather than a single design ; that permits the Genetic algorithm to keep a multipoint observation on several locations within the solution space simultaneously, resulting in a high possibility of discovering the global optimum [8].. The flow chart that shows the working of genetic algorithm is given below in Figure 2.

3.3 Particle swarm optimization

Kennedy and Eberhart (1995) developed particle swarm optimization technique (PSO) for optimal solutions. Particle Swarm Optimization technique has an initialization and an evolution phase. In initialization phase, population is initialized using randomly distributed particles within the search, whereas in evolution phase particle adjust their positions according to the most successful particles within the space till the termination of the met criteria.

Virtual particles move in D-dimensional space in the PSO model. Every particle keeps best solutions in its path named as pbest and best location met by the swarm named as gbest. The next movement of individual particle is controlled by a velocity vector and influenced by gbest and pbest.

The position of the ith particle in the search space is indicated as $k_i = k_{i1}, k_{i2}, \ldots, k_{iD}$ and velocity is indicated as $V_i = v_{i1}, v_{i2}, \ldots, v_{iD}$. The position and velocity are updated according to equation 6 and 7 given below

$$v_i^{t+1} = w * v_i^t + c_1 rand_1 (pbest_i - k_i^t) + c_2 rand_2 (gbest_i - k_i^t) \quad (6)$$

$$k_i^{t+1} = k_i^t + v_i^{t+1} \quad (7)$$

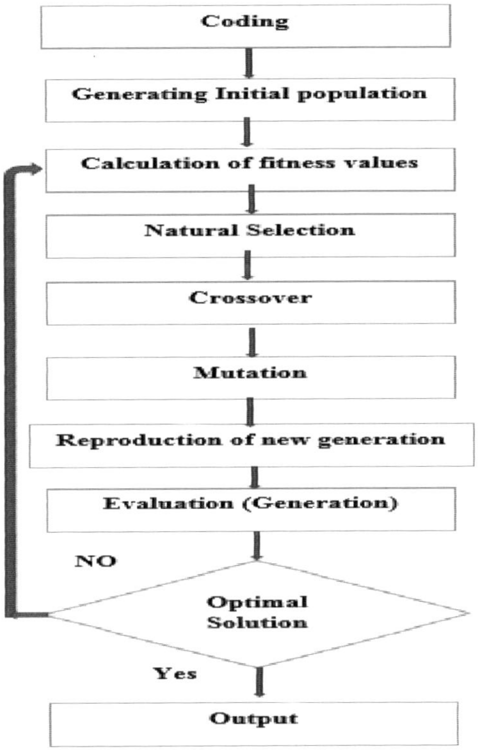

Figure 2: Flow chart of Genetic Algorithm

Where $k^{min} < k_i^{t+1} < k^{max}$, i=(1,2,3….swarm/population size);t=number of reproduction steps; v_i^t =speed vector of k particle in ith reproduction; w is inertial weight; c_1 and c_2 are learning rates; $rand_1$ and $rand_2$ are independent uniformly distributed random variables within [0,1] range; $pbest_i$ and $gbest_i$ are best solution reached by particle k and swarm respectively.

4 Results and discussion

Inflow and outflow data of the Mangla reservoir since 1960 to 2010 was collected from Water and Power Authority of Pakistan. Log three Pearson technique had been applied at this data to determine the year with maximum, average and minimum flows in the Mangla watershed. The year with maximum flows named as rainy year and with average flows named as average rainfall year. Similarly, the year observed with minimum flows named as Dry year. The results obtained after the application of different algorithms at these data series are presented in Table 1 and for PSO are shown in figure discussed below.

4.1 Genetic algorithm

Results obtained after the application of genetic algorithm at rainy, average and dry year data series are presented into Table 1 for energy output and time of execution of algorithm. Results depict that upto $59*10^8$ kWh electricity could be produced by using GA for rainy year and time required for this execution is about 16.82 seconds. Results depict that we can produce maximum electricity for rainy, average and dry years by using GA. Table 1 presents that time of execution is more than PSO for rainy, average and dry year. Results exhibits that electricity production could be increase if we use the flows according to GA as it gives much more amount of electricity than traditional particle swarm optimization methods.

4.2 Particle swarm optimization

Results obtained after the application of particle swarm optimization technique at rainy, average and dry years are presented into Table 1 and for rainy and average year in Figure 3. Results depict that electricity production using particle swarm optimization is lower than GA but time of execution for PSO is much lesser than GA. Moreover, GA and PSO produced more electricity than traditional method. Figure 3 presents upto $53*10^8$ kWh of electricity using PSO with outflow and water level shown for rainy year. It is also evident from the Figure 3 that upto $33.7*10^8$ kWh electricity can be produced from average year with water release and water level given in the Figure 3.

Table 1: Electricity production using different algorithms

Algoriths	**Energy Output (10^8 kWh)**				**Execution Time (s)**			
	Rainy Year	Average Rainfall year	Dry Year	Average	Rainy Year	Average Rainfall year	Dry Year	Ave rage
Traditional Method	47	25	20	30.66				
GA	59	43.8	38.2	47	16.82	16.76	16.68	16.8
PSO	52	33.7	28.3	38	3.88	3.84	3.81	3.84

5 Conclusion

The present study focusses on the optimal electricity generation from the Mangla reservoir in Pakistan. A mathematical model has been developed for the Mangla hydropower station and particle swarm and genetic algorithm optimization techniques were applied at this model for optimal electricity generation. Results revealed that Genetic Algorithm can produce maximum electricity generation as compare to Particle swarm optimization, but the time of execution of particle swarm optimization is much lesser than the Genetic algorithm. Results revealed that the electricity production increases with the application of genetic algorithm and particle swarm optimization techniques at proposed mathematical model than traditional techniques. Results are evident of up to $59*10^9$ kWh electricity production with optimal flows that is more than $47*10^8$ kWh, that we are getting using traditional method in Mangla hydropower station. Results revealed that by managing our discharge and water level equivalent to optimal water level, we can produce maximum electricity. More benefits can be obtained in the form of power production with the application of Genetic algorithm at mathematical model than PSO for Mangla hydropower station.

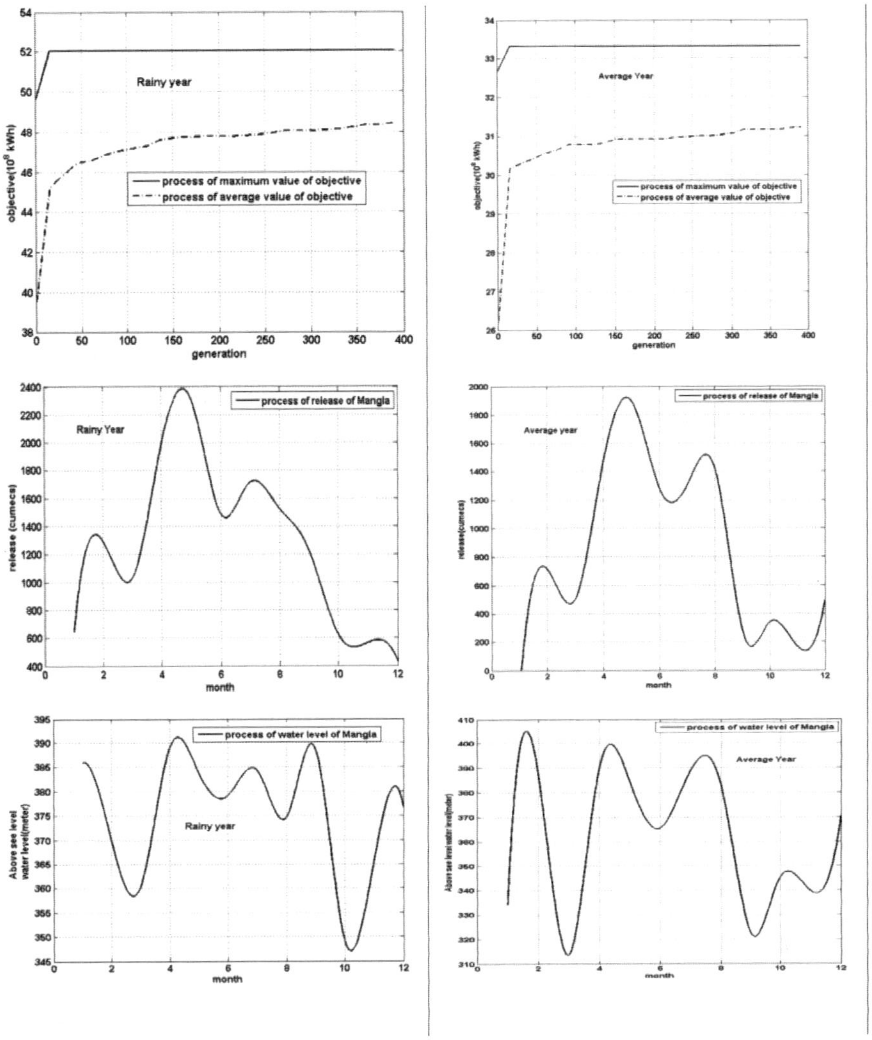

Figure 3: Results of PSO for rainy and average year for Electricity production, Water level and Water release

Acknowledgements

This work is supported by the National Key Research and Development Program '' Research and Development of Green and High Efficient water saving Irrigation Equipment's for Typical Rural Areas in Northwest China' No 2016 YFC 0400202.

References

1. Hejazi, M. & Cai, X. Building more realistic reservoir optimization models using data mining–A case study of Shelbyville Reservoir. *Adv. W. Resour.* **34**,701–717 (2011).

2. Sniẹdovich, M. Reliability-constrained reservoir control problems: 1. Methodological issues. *Water Resour. Res.* ***15*(6)**, 1574-1582. (1979).

3. Su, S. & Deininger, R. Modeling the regulation of Lake Superior under uncertainty of future water supplies. *Water Resour. Res.* ***10*****(1),** 11-25 (1974).

4. Mousavi, S. & Karamouz, M. Fuzzy-state stochastic dynamic programming for reservoir operation. *J. water Resour.* **130(6)**, 460-470 (2004).

5. Huang, W. & Yuan, L. A drought early warning system on real-time multireservoir operations. *Water Resour. Res.* **40(6)** (2004).

6. Ganji, A., Karamouz, M. & Khalili, D. Development of stochastic dynamic Nash game model for reservoir operation II. The value of players' information availability and cooperative behaviors. *Adv. Water Resour.* **30(3)**, 528-542 (2007).

7. Kucukmehmetoglu, M. A game theoretic approach to assess the impacts of major investments on transboundary water resources: the case of the Euphrates and Tigris. *Water Resour. Manag.* **23(15)**, 3069-3099 (2009).

8. Goldberg, D. Genetic algorithms in search, optimization, and machine learning, 1989. *Read. Addison-Wesley* (1989).

9. Chang, F., Chen, L. & Chang, L. Optimizing the reservoir operating rule curves by genetic algorithms. *Hydrol. Process.* **19(11),** 2277-2289 (2005).

10. Jian-Xia, C., Qiang, H. & Yi-Min, W. Genetic algorithms for optimal reservoir dispatching. *Water Resour. Manag.* **19(4)**, 321-331 (2005).

11. Oliveira, R. & Loucks, D. Operating rules for multireservoir systems. *Water Resour. Res.* **33(4)**, 839-852 (1997).

12. Wardlaw, R. & Sharif, M. Evaluation of genetic algorithms for optimal reservoir system operation. *J. water Resour. Plan.* **125(1)**, 25-33 (1999).

13. Sharif, M. & Wardlaw, R. Multireservoir systems optimization using genetic algorithms: case study. *J. Comput. Civ. Eng.* **14(4)**, 255-263 (2000).

14. Zyl, J. Van, Savic, D. & Walters, G. Operational optimization of water distribution systems using a hybrid genetic algorithm. *J. water Resour.* **130(2)**, 160-170 (2004).

15. Wang, K., Chang, L. & Chang, F. Multi-tier interactive genetic algorithms for the optimization of long-term reservoir operation. *Adv. Water Resour.* **34(10)**, 1343-1351(2011).

16. Kennedy, J. & Eberhart, R. Particle swarm optimization. *1995. Proceedings., IEEE Int.*pp. 39-43 (1995).

17. Rahi, O., Chandel, A. & Sharma, M. Optimization of hydro power plant design by particle swarm optimization (PSO). *Procedia Eng.* **30,** 418-425 (2012).

AUTHOR INDEX